T0199897

SCIENCE AND THE CREATIVE IMAGINATION IN LATIN AMERICA

Science and the Creative Imagination in Latin America

Edited by

Evelyn Fishburn and Eduardo L. Ortiz

Institute for the Study of the Americas
Senate House, Malet Street, London WC1E 7HU
Web: www.sas.ac.uk/americas

British Library Cataloguing-in-Publication Data
A catalogue record for this book is available
from the British Library

ISBN   1 900039 61 3

INSTITUTE FOR THE STUDY OF THE
A M E R I C A S
UNIVERSITY OF LONDON · SCHOOL OF ADVANCED STUDY

Institute for the Study of the Americas
Senate House
Malet Street
London WC1E 7HU

Telephone: 020 7862 8870
Fax: 020 7862 8886

Email: americas@sas.ac.uk
Web: www.sas.ac.uk/americas

TABLE OF CONTENTS

NOTES ON CONTRIBUTORS

Dame Gillian Beer is King Edward VII Professor of English Literature Emerita. She is a former president of Clare Hall, Cambridge, and her books include *Darwin's Plots: Evolutionary Narrative in Darwin, George Eliot and Nineteenth-Century Fiction* (Cambridge, 2000) and *Open Fields: Science in Cultural Encounter* (Oxford, 1999). She was vice-president of the British Academy from 1994 to 1996.

Claudio Canaparo is Reader in Latin American Studies in the School of Modern Languages at the University of Exeter. He is interested in sociology of knowledge and philosophy of culture in Latin America, and is founder of the Interdisciplinary Research Centre for Latin American Studies (IRCLAS). His books include *Ciencia y escritura* (Buenos Aires, 2004) and *Muerte y transfiguración de la cultura rioplatense* (forthcoming).

Evelyn Fishburn is Emeritus Professor of Latin American Literary Studies at London Metropolitan University and Honorary Senior Research Fellow at University College London. She has published numerous articles on Borges in particular and Latin American literature generally. She is the editor of *Short Fiction by Spanish-American Women* (Manchester, 1998) and *Borges and Europe Revisited* (London, 1998).

Norma S. Horenstein was a Lecturer in the Faculty of Humanities and Philosophy at the Universidad Nacional de Córdoba. She was a founder-member of the Asociación de Filosofía e Historia de la Ciencia del Cono Sur. She co-edited, with Leticia Minhot and Hernán Severgnini, *Epistemología e historia de la ciencia. Selección de trabajos de las XII Jornadas* (Córdoba, 2002). She died in 2002.

Geoffrey Kantaris is University Senior Lecturer in the Department of Spanish and Portuguese at St Catharine's College, University of Cambridge. He specialises in Latin American cinema and questions of modernity and postmodernity in Latin-American urban culture. He is the author of *The Subversive Psyche: Contemporary Women's Narrative from Argentina and Uruguay* (Oxford, 1996).

Alejandro Kaufman is Professor of Communication Theory and Subjectivity in the Universidad de Buenos Aires and the Universidad de

Quilmes. He is the author of *Lyotard, Heidegger y los judíos* (Buenos Aires, 1995) and co-author of *Itinerarios de la modernidad* (Buenos Aires, 1999).

Sylvia Molloy is Albert Schweitzer Professor in the Humanities at New York University, and she is a former president of the Modern Language Association. Her books include *Signs of Borges* (Durham, NC, and London, 1994) and *At Face Value: Autobiographical Writing in Spanish America* (Cambridge and New York, 1991).

Patricia Murray is a Senior Lecturer in the Department of Humanities, Arts and Languages at London Metropolitan University. She is a specialist in postcolonial literatures and critical theory, and is the co-editor, with Ashok Bery, of *Comparing Postcolonial Literatures* (Basingstoke, 2000).

Eduardo L. Ortiz is Emeritus Professor of Mathematics and History of Mathematics, and Senior Research Investigator, at Imperial College London. He specialises in the history of science and mathematics in the nineteenth century. His latest work is *Mathematics and Social Utopias in France, 1810–1850* (with S. Altmann) (forthcoming).

Alicia Rivero is Associate Professor of Spanish American and Comparative Literatures in the Department of Romance Languages at the University of North Carolina. She has written widely on Spanish American and comparative literatures, the relationship between literature and science and other aspects of cultural studies. She is the author of *Autor/lector: Huidobro, Borges, Fuentes, y Sarduy* (Detroit, 1991) and the editor of *Between the Self and the Void: Essays in Honor of Severo Sarduy* (Boulder, 1998).

William Rowe is Anniversary Professor of Poetics in the Department of Spanish and the School of English and Humanities at Birkbeck, University of London. He has written widely on cultural history and modern poetry and poetics in the Americas. His books include (co-authored by Vivian Schelling) *Memory and Modernity: Popular Culture in Latin America* (London, 1991) and *Poets of Contemporary Latin America: History and the Inner Life* (Oxford, 2000).

Benigno Trigo is Associate Professor in the Department of Spanish and Portuguese at Vanderbilt University. He has written on literature's relation to scientific discourse in *Subjects of Crisis: Race and Gender as Disease in Latin America* (Middletown, CT, 2000). He is also the editor of *Foucault and Latin America: Deployments and Appropriations of Discursive Analysis* (Oxford, 2001), and co-author with Kelly Oliver of *Noir Anxiety* (Minneapolis, 2003).

Introduction

Evelyn Fishburn and Eduardo L. Ortiz

One of the most exciting recent developments in interdisciplinary studies has been the dismantling of the oppositional stance traditionally occupied by the humanities and the sciences. According to legend, Bertrand Russell's grandmother would tease him about his philosophical interests by chanting 'What is mind? Never matter! What is matter? Never mind!' The joke, presumably, lay in speaking about such dissimilar, unbridgeable notions in the same context, whilst naturally emphasising their difference. But today's postmodern and post-Einsteinian reader might view the aphorism differently, seeing in it a circularity that suggests a continuum rather than a break. For the notion of 'Two Cultures', as coined by C.P. Snow in his 1959 Rede Lecture has clearly become outmoded through the erosion of boundaries that has gradually taken place not only within modern science and literature but also between modern science and literature. Snow's contention, though a major influence on western thought, was both exaggerated and polemical even at the time he made it, and it was never as pertinent in Latin America, where much closer links have always existed between the humanities and the sciences. Science has traditionally been less secluded in its history: the effects of positivism in Argentina, the *Científicos* in Mexico, the motto *Ordem e progresso* on the Brazilian flag, the prominence of science in the educational system and the many statesmen and indeed writers who were also scientists all point to a degree of integration that confutes Snow's model.

The present volume considers some of the connections and continuities in the relations between the humanities and the sciences in a Latin American context. The geographical emphasis is important given the prominent role that science played in the formation of the nation states in Latin America and its strong presence in the cultural output of the area. And yet, it remains an issue that has not been addressed systematically, although there have been important isolated studies of the influence of scientific ideas upon specific writers.

Most of the chapters in the present volume are concerned with fictional narratives and scientific discourses. In exploring possible links between them, related modes of perception and expression are highlighted, though neither in a causal nor in a conciliatory way but rather as expressions of common anxieties.

Questions of consent, resistance and ideology in both fields are considered, as is the semiotic complexity of these productions. The historical study of interplays between science and the novel not only allows us to identify what was expected to be believed at a given point in time, but offers some insights into how this belief was helped and sustained. It also assists us in understanding the interplay between what the individual understood of himself or herself and the surrounding world of science.

Professor Dame Gillian Beer, whose essay begins the collection, was amongst the first to ask the key questions regarding the place of science in our cultural imaginary. Questions such as: How do groups learn from each other when they cannot interpret each other's signs? How are scientific ideas realised in the minds of the scientists and, how are they understood, and realised, outside the scientific community?

Postcolonial studies have made us conscious of the difficulties in interpreting different sign-systems across cultures, but have shown too the gains to be had from an outsider's, or marginal perspective. Borges's increasingly important essay, *El escritor argentino y la tradición*, argues for the special perceptions of those who are not totally immersed in, and therefore bound by, one culture alone, a notion that can be translated to the topic at hand. In the long wake of the Sokal hoax it might seem foolhardy, to say the least, to venture into a border-crossing exercise over such seemingly diverse fields as the sciences and the humanities, but we believe that 'misreadings', or 'creative misreadings' to abuse, perhaps, Bloom's expression can have an illuminating effect.

Although several approaches have been taken, the main area of enquiry has been socio-cultural: nineteenth-century travel writings, cartography, portraiture and other visual imagery, with particular emphasis on their inscription in the national cultural imaginary; the protean use of scientific arguments in the construction of social mythologies; discussions of the overlaps of quantum physics and quantum poetics; simulacra and the relation of science and technology to its fictional representation; and, turning the tables, the effect of writing about science upon science.

Gillian Beer's introductory chapter traces the conceptual developments leading to Darwin's theory of evolution, and to his travels inland and around the shores of South America. The continent, with its diversity, excess, contradictions and above all with the evidence of ongoing change, became for him the vantage point from which to re-think the normative assumptions of his own British culture.

Diversity presented itself to Darwin in the form of organic life as well as social structures; it showed him that problems may have several possible solutions, widening his concept of causality beyond the restricted sense of the physical sciences of his time. On a more personal level, the voyage allowed him to discover in himself, and also to express, an exceptional capacity for creativity, which he showed in his life-long analysis of change and diversity.

Beer points out that change requires extinction as much as continuity, forgetting as well as memory. This thought is related to Darwin, who worked from his notes and collected memorabilia on a concept of the universe which extended far beyond human memory; as a geologist he was conscious of a universe which predated the existence of man, and greatly exceeded the parameters of human thought and memory. Memory, therefore, became an important though insufficient tool to develop the long history of the earth, and of life on it. Lamarck had given memory, the past experiences of species, a place in his view of the larger scheme of evolution but Darwin now rejected this idea postulating that species do not retain their life experience.

Interestingly, poetry played an important role in the development of Darwin's ideas: he drew freely from his literary readings, and moved seemingly unencumbered from scientific and non-scientific thought. A copy of *Paradise Lost* was his constant companion on the voyage and he found inspiration for his ideas on the mechanism of evolution from its account of creation. In her analysis of Darwin's *Notebooks*, Beer notes a shared linguistic discourse between literary and scientific thought. For Darwin, literary techniques such as metaphor and analogy are all-important in freeing the imagination to perceive new insights and make new connections. A new language was needed to move into new 'zones of thinking' and Darwin's writings provide us with an early version of it. Body and the surrounding world were given a new reading.

The evolution of scientific ideas through encounter and travel writing is also examined by Trigo and Molloy. Trigo discusses global views on Latin America through the narratives and works of three European travellers writing in the first half of the nineteenth century. They are the Prussian Humboldt, the Frenchman Mollien and the Italian Codazzi. Trigo tests their views, expressed in narratives and extensive geographical works, on the concrete case of Colombia.

Arguing against the perception that Humboldt reduced America to a landscape, marginalising its inhabitants, Trigo focuses on his romantic notion of union between subject and cosmos, and on his view that Latin America was a continent on the brink of revolution, where problems related to time, space, race, change and disintegration were highlighted. These views became dominant during the patrician period of Colombian history.

Mollien, on the other hand, emphasised biology over environment, in accordance with a new historical perspective in vogue at the time, and offered a less optimistic view. He gave disease, or degeneration in all its forms, a more dramatic dimension. Codazzi's Chorography Commission is seen as a project through which the young *letrados*, a new formulation of the old patrician society, reacted against their predecessors' views on Latin America.

In time, Colombia ceased to be regarded as a country of the future, in which technology would help man to dominate nature. A more realistic view, in which the position of slavery is reconsidered, and the huge impact of civil wars is taken into account, prevailed. The old perception that sustained views which can be related to the ideology of enlightened despotism, and which held that change could be swift, perhaps violent, and supported from outside, gave way to a more federally-oriented perspective, tending to put the accent on communication and exchange. With these new perceptions, a more fluid flow of population, merchandise, and also of cultures, was incorporated to the imaginary on Colombia.

Trigo uses illustrations of *cargueros*, the persons who carried a passenger on their shoulders, to illuminate the different perspectives of the Indian, those of African descent and the Europeanised *criollos*.

In Sylvia Molloy's contribution science is shown to be a far from objective pursuit. Her essay concerns the formation of national scientific collections and exhibits in late nineteenth-century Argentina, largely with materials collected during and after what is polemically known as 'la conquista del desierto'. This 'conquest of the desert', though linked inevitably in Argentine consciousness with heroism and patriotism, was not confined to military exploits; scientists, archaeologists, botanists, geographers and anthropologists also travelled south, especially to Patagonia. On the whole, they were less intent on pushing the Indian away from the city than on bringing him back in pieces as an object of study and an item for a collection.

Molloy argues that this appropriation of artefacts, of human remains, of botanical and mineral specimens designed to signify the nation's wealth was another form of patriotic expansionism, which she terms a *pro patria* endeavour. She illustrates this argument with an analysis of a travel book, *Viaje a la Patagonia austral*, by Francisco P. Moreno (1852–1919), founder and then director of the Museo de Ciencias Naturales de La Plata. The author was a natural scientist who, like Darwin before him, but with different aims, travelled south to study the native Indians, amassing an array of material of local interest. Moreno constructs himself and his travels as heroic and patriotic. Molloy's assessment is more nuanced: she notes that as an anthropologist he wishes to study the Indians' customs and how they can best be integrated into the nation's workforce; as an archaeologist, however, he uses the Indian as an inert object of curiosity, a museum exhibit, but with a hidden purpose. This is to reinforce the sense of the nation's autochthonous past in order to counterbalance the cultural threat of massive immigration.

Naming is a form of possession. During his travel in the region Moreno claimed landmarks for his country by the expedient of giving them names that reflected Argentine interests. The first part of the title of Molloy's essay, 'National Parts and Unnatural Others: a Reflection on Patrimony at the Turn of the Nineteenth Century' refers, to the conversion of natural parts, whether of human remains or geographical sites, to national parts through nomenclature. The 'unnatural others' are those that have not been integrated, or naturalised, into the country's symbolic imaginary. Molloy quotes from a short story by the political scientist Octavio Bunge. It concerns a siren who inspires the poetic and erotic imagination of her captors, but is ultimately returned to the seas because she is different: the discarded siren, an exotic, unnatural Other, does not fit into the scientific and patriotic project of the moment.

An appraisal of the use of scientific ideas in the construction of the nation as expressed in fictional and non-fictional texts links Molloy's chapter with the contributions by Ortiz, Kauffman and Canaparo. Eduardo L. Ortiz's chapter identifies a key moment in the transition from realism to the fantastic in the Argentine literature of the 1870s. Relying heavily on his recent research into the history of science in Argentina, a new interpretation is proposed for Eduardo L. Holmberg's work *Viaje maravilloso del Señor Nic Nac*, published in 1875–76. This important, and largely neglected, novel has been regarded as a purely fantastic story, with totally fictional situations and characters. In Ortiz's reading, its supposedly fantastic element is shown to have a much closer relationship with realism than previously thought.

Nic Nac is a novel based on a real historical episode, here referred to as the 'Córdoba Six'. This was a controversy between by a group of German professors hired by the Argentine government to encourage the development of science in the city of Córdoba, seat of one of the oldest universities in the Americas, and Germán Burmeister, another leading German scientist who ran, rather autocratically, the Natural History Museum in Buenos Aires. The affair ended, dramatically, in a Pyrrhic victory for Burmeister.

This episode took place at the time of the foundation of one of the leading scientific institutions of Argentina, the National Academy of Sciences. The characters in Holmberg's novel are clearly fictional, but are constructed from real actors and situations taken from this most interesting episode in the early history of the insertion of modern science in Argentina.

Kaufman's chapter opens with an overview of the question of the two, or perhaps three cultures (in Snow's definition); it then considers attempts to formulate a unitary frame for culture, and ends with a discussion of the interplay between the literary and scientific subcultures, stressing developments in this direction in the second half of the twentieth century.

His discussion of Martínez Estrada's work begins with the observation that the author of *Radiografía de la pampa* has metaphorically and systematically used scientific language when describing socio-historical events. Kaufman reminds us that the same conceptual strategy can be traced in his *Análisis funcional de la cultura*, published some 30 years later. In this last work, analysing Sarmiento's dichotomy between civilisation and barbarism, Martínez Estrada points out that, far from the one having defeated the other, civilisation and barbarism have re-emerged united, hidden under the trappings of prosperity and mechanical and cultural advances. He also discusses the use of a scientific vocabulary in some of Martínez Estrada's specific works.

Kaufman recalls that when Martínez Estrada was invited by the Universidad del Sur, in Bahía Blanca, to deliver his famous university lecture of 1959, at a critical time in the history of culture in Argentina, when the state was considering unloading popular and higher culture from traditional national responsibilities, he asserted his preference for the temple of Montaigne, Thoreau and Nietzche, rather than the ironmongeries of the idols of technology. In the same oration, he stressed that a reform of university teaching could only be constructed on the basis of a sound philosophical conception of culture.

Nature: A Weekly Illustrated Journal of Science is a British journal created in 1869 to report on the progress of the representation of nature in man, which, as stated by T.H. Huxley in the opening words of this journal, is what we call science. Since its inception it has been a model for scientific journals throughout the world. When the Institución Libre de Enseñanza commenced publication of its own influential *Boletín* in Spain in 1877, it adopted the same size, the same paper and the same internal structure as *Nature*, with original articles, national and international news and book reviews.

The British journal's impact can be detected in a number of other publications in the Spanish-speaking world. Using sociological analytical tools, Canaparo discusses a new attempt, in the long century that separates us from its first recreation in this language in Madrid. This is the journal *Redes*, founded in 1994 as an effort to advance social studies of science and technology in Argentina. The journal anchored these studies to recent Anglo-Saxon trends and in a detailed dissection of this journal, Canaparo detects fluctuations in the choice of possible theoretical paths in its approach to science and technology. Stressing the development of a 'ciencia propia', it initially rejected the possibility of a larger 'Latin American model', while considering its merits later and, finally, adopting a more internationalist approach, offered its readers works of leading European and American theoreticians, such as Bourdieu, Latour and others.

Canaparo's interest in *Redes* is to research the views on science and technology held by its editorial group. He discusses the institutional associations of this journal, intersections between the communities of its editors, staff, contributors and reviewers, as well as its deployment of academic elements and its use of journalistic tools. He also offers a detailed analysis of a fracture in *Redes*' focus: an analysis of science as a written phenomenon, as social studies of science, but also a concern with current local scientific policies. Canaparo points out that the journal's concern with field studies of scientific communities is significantly greater than with the epistemological debate on how scientificity is constructed.

The representation of science in works of fiction (and film) is addressed from a variety of angles in the remaining essays in this volume. This is initially considered by Norma Silvia Horenstein, who died in 2002, and whom we remember with sadness and affection. Her angle on the interplay of science and the creative imagination is to focus on the translatability of ideas from one era to another and one discourse into another. She identifies traces of the Pythagorean–Platonic and Newtonian scientific traditions in the definition of cosmogonic canons alluded to in Canto XXX of José Hernández's epic poem *Martín Fierro*. As Borges has commented, this is a privileged moment in the poem, the mise-en-abime of a *payada* that is inside the overall *payada* of the poem, and as such it merits special and different attention. The personal and socio-political preoccupations of *Martín Fierro* are interrupted in a song contest discussing ontological problems related to mathematics and physics. The main issues of scientific sign are: 1) the notions of quantity and measure, concepts of Pythagorean inspiration in Greek scientific thought; and 2) the idea of weight, understood in terms of Newtonian ideas on gravitation.

Horenstein offers a detailed explanation of the scientific ideas which she detects lie behind the succinct verses of the *payada*, making the point that the poet must have been familiar not simply with the relevant general ideas but with their particular development and discussion. Following the threads of Astrada and Martínez Estrada in their discussion of José Hernandez's epic poem, Horenstein addresses the topic of José Hernandez's education. He did not go to university and had no formal education beyond elementary school, although he received some instruction when he joined the army at the age of 19. This leads Horenstein to enquire how Hernández could have acquired such a profound understanding of complex concepts of mathematics and physics. The subject of Hernández's education has not been fully explored, but Horenstein puts forward the argument that at that time in Argentina access to scientific ideas was not limited to the elite intellectual environment of secondary schools or universities.

Alicia Rivero discusses correlations between the findings of quantum physics (the notion of a discontinuous universe as classically defined without cause and effect) and the development of twentieth-century narrative techniques. This is a topic that has so far been addressed mainly with reference to Anglo-European texts; yet quantum physics and particularly Heisenberg's uncertainty principle, often perceived as stating that all that we have access to is an observer-created reality, are assumptions of specific interest to the study of Spanish-American texts. Cultural considerations are obviously fundamental, and Rivero's argument *in nuce* is that this is essentially a culturally hybrid continent, and that the observer as much as the observed are affected by this hybridity. The science that inflects the literature examined is not simply a borrowing or imitation of western science but the product of a centuries' long process of transculturation.

Rivero offers a brief overview of twentieth-century physics in Spanish America, followed by an introductory explanation of relevant aspects of quantum mechanics. The emphasis is on the constructedness of the scientists' epistemological interpretation of data, a concept which is echoed in literary studies in terms of metafiction and reception theory.

The second part of this chapter focuses more intensely upon literary texts. In her reading of 'quantum fiction', fiction which questions realist suppositions as does Heisenberg, Rivero draws parallels between the 'interference' of the physicist and that of the (active) reader. She probes the way in which quantum ontology and epistemology are reflected in some of the themes and strategies used by modern Spanish American writers who actively involve the reader as a textual participant. Different aspects of this topic are briefly traced in the poetry of Cardenal, in an essay by Vasconcelos and in the writings of Borges, Cortázar and Volpi, ending with a more detailed examination of Fuentes's *Cristóbal Nonato*. In this polyphonic novel, Cristóbal, the principal narrator who explicitly mentions Heisenberg, affirms that observing phenomena changes them. Rivero shows us that Cristóbal's statement implicates both the narrator and the reader in the creation of textual reality in a manner akin to the tenets of quantum physics, especially those espoused by Heisenberg.

Patricia Murray looks at the relationship between science and scientific perspectives and the theme of postcoloniality in García Márquez's *Cien años de soledad*. Somewhat against the grain of dominant criticism on the novel, she reads García Márquez's masterpiece of 'lo real maravilloso' as a clarion call to the future (subversive and optimistic) rather than as a nostalgic critique of the past. The failure of the Buendías is contrasted with the richly diasporic and multilingual contact zone of Macondo, a place where the act of memory central to García Márquez's scientific concerns becomes possible. With reference to its status as quantum fiction (and to the guiding hand of Borges) Murray is keen to reiterate the complexity of the novel's form, the complementary and partial nature of the text we are reading and the flexibility of interpretation it demands.

Though often captivated by the wonder of science, and inspired by all manner of scientific enquiry, the novel is concerned with notions of where and how science gets produced, how it gets remembered, used and manipulated. The boundaries we draw in defining what is, and what is not, a scientific perspective are often linked to histories of cultural dominance and Murray contends that the narrative of scientific progress which takes place in the novel (culminating in the predatory and capitalistic enterprise of the banana company) is sharply critical of a modernity that instigates an amnesiac fate. It is the cross-cultural figure of Melquíades, a scientist-seer who retains multiple histories of scientific and philosophical exchange, who operates as the key subaltern influence and source of learning in the novel. He acts as a vivid reminder of the exclusions of modernity and of the eclipsed perspectives that post-colonialism must learn to remember.

Murray argues that for García Márquez in 1967 (as indeed for us today) postcoloniality was still an anticipatory discourse, a future that has still to be written, but she attempts to show how a version of postcoloniality emerges out of the various scientific enquiries that take place in the novel.

The gradual replacement of a real empire by its simulacrum, a map, is ludically implied by Borges in *Del rigor en la ciencia*; Baudrillard uses the Borgesian trope to introduce his own discussion of simulation and simulacra arguing that in the present era of technological simulation what matters is not the real (as conceived in nineteenth-century scientific terms), but its representation. This idea underpins Kantaris's essay exploring the presentation of science not as traditionally understood, as the study of the physical and natural world, but as a discourse by which the simulation of this world is produced. The feedback loop between new technologies of representation and the fictional representation of technology is here examined in three texts, *La invención de Morel* (novella) by Adolfo Bioy Casares, *Hombre mirando al sudeste* (film) by Eliseo Subiela and *La ciudad ausente* (novel) by Ricardo Piglia.

Kantaris traces the impact of technology upon the cultural imagination in Argentine modernity to Arlt, whose protagonist (Erdosain) dreams of changing the real through his technological innovations. Such innovations are seen at work in *La invención de Morel*, in which a fugitive has fallen in love with a simulacrum, the hologram produced by a machine that ultimately erodes and destroys the matter it has recorded. The text of this work of science fiction is made up of the writings of three unreliable narrators, and this affirmation of its own artifice seems to exemplify the point being argued. Subiela's film *Hombre mirando al sudeste* shows the difficult interaction between an outer-space patient and his psychiatrist. The relationship between hallucination and reality is seriously problematised in this film, while its setting in a psychiatric clinic, a metaphor for the state, serves to introduce the issue of control. Strategies of (state) control over proliferating simulacra are further explored through a story-telling machine enclosed in an imaginary museum in Piglia's futurist fantasy novel *La ciudad ausente*. Kantaris's conclusion is not overwhelmingly pessimistic: simulation can evoke strategies of resistance which often lead to a return to fundamentals.

In his discussion of the relationship between the sciences and the humanities William Rowe points out the difficulty of studying the one in terms of the other, if only because what are considered 'facts' differs according to each discipline and to the time and place of enunciation. And in poetry, tone plays a significant part, beyond the conceptual. These issues are at the forefront of Rowe's reading of *Fin desierto* by the Peruvian poet and linguistic theoretician, Mario Montalbetti. Montalbetti's work is a critique of language as 'the institution *par excellence* through which we make sense of reality'. Instead of assuming, as cultural criticism seems to do, that all experience can be written, his poetry draws attention to what cannot be written, to the essential inexpressibility of physical reality.

In an attempt to present the elements of a theory of language, Rowe ponders the effect of the desert upon linear orientation, upon questions of scale (measurement), rhythm and texture, exploring the consequences of particular knowledge for language.

The desert becomes an Archimedian point from which to view the conceptual world we inhabit and that which is absent from it. Montalbetti's poetic work discussed here is directly concerned with the Peruvian desert: its immensity, and its nearby beaches, where immense extensions of sand border equally immense extensions of water. He also considers the desert as the repository of pain and genocide, through ancient and abandoned burials:

> *Solo trapos*
> *y cráneos de los muertos, nos anuncian*
> *que bajo estas arenas*
> *sembraron en manada a nuestros padres.*

Fin desierto uses the metaphor of crossing the desert as the condition of writing. It claims concern not with the experience of the desert but with its minimalist physical properties, which it argues are related to the properties of narrative. Meaning resides and is dependent upon them. This is reflected to an extent by a screenfold layout, empty spaces and the unusual typographical design of Montalbetti's book, and should be considered in terms of the essay's epigraph, 'Art does not illustrate philosophy, *it comes before it*'. Rowe uses Montalbetti's book as an artefact to explore, in depth, poetry and its relations to philosophy and science; the relation or lack of relation between writing and knowledge is an underlying concern. He focuses in particular on the 'ontological commitment' (qué es lo que existe y qué es lo que no existe) displayed by Montalbetti in *Fin desierto*, and relates it to the epistemological concerns of philosopher José Carlos Ballón regarding the interplay of science and philosophy. Other wide-ranging references related to this discussion are also invoked. Rowe concludes that Montalbetti and Ballón can also be related in other instances, for example, in their treatment of knowledge and misfortune as inseparable qualities, since one is 'as objective and universal as the other'.

As is evident from the foregoing, the studies in this volume cover a wide variety of topics considered from different perspectives, yet they are bound by a shared concern, which is the place of science in the Latin American cultural imaginary. As editors we have opted for an open-ended approach, since our interest was in initiating debates rather than drawing overarching conclusions. Our sincere hope is that this compilation will serve not only to expand perceptions of the past but inspire new modes of thinking and new directions for future research.

We wish to thank the Institute for the Study of the Americas, School of Advanced Study, and particularly its director, Professor James Dunkerley, for encouragement and constant support while this work was developed and for giving us generous facilities to organise a workshop on Science and Literature in which the contributions included in this volume were collectively discussed.

1

Darwin in South America: Geology, Imagination and Encounter

Dame Gillian Beer

Imagine Darwin, a young man of 30 in 1839, at home in London. By then he has half-trained to be a medical doctor and, like the poet Keats ten years before him, withdrawn from the course partly because of the pain medics were bound to inflict before the coming of anaesthesia; he has spent three years at Cambridge doing, like most of his contemporaries, a general degree and foreseeing a parsonage as his future. Then suddenly just before he is 23 (the age of a graduate student now) he has set off round the world on 27 December 1831 on the *Beagle*. The next five years transform his life and his imagination — and the first three of them are spent almost entirely in the countries of South America: on inland treks across the plains, in tropical forests, around the shores, and visiting islands and land clusters including the Falklands, Tierra del Fuego and eventually the Galapagos. During that time he encounters revolution, earthquake, the slaughter of Indian populations, slavery, beetles and polyps, boredom, months in the saddle, ecstasy, indigenous peoples and endless onward travel, always leaving behind much of what he needs for evidence and having to learn to evoke as thought and image what has become absent and irretrievable as sense experience.

He returns in 1837, and by 1839 he has recorded, secretly, in his notebooks the crucial concepts of his evolutionary ideas. The notebooks are full of glimpses, asides, essays at thought:

> The tree of life should perhaps be called the coral of life, base of branches dead; so that passages cannot be seen.

> When one sees nipple on man's breast, one does not say some use, but sex not having been determined. — so with useless wings under elytra of beetles. — born from beetles with wings & modified — if simple creation, surely would have been born without them.[1]

1   Howard E. Gruber (1974) *Darwin on Man: A Psychological Study of Scientific Creativity together with Darwin's Early and Unpublished Notebooks*, transcribed and annotated by Paul H. Barrett (London: Wildwood House), pp. 442 and 445.

He works with metaphor (the conventional upward–reaching tree of life becomes the downward-collapsing coral, its base crumbled away beyond evidence) and with analogy: male nipples and useless beetle wings. Both the examples above explore time and waste, until meaning is stretched to encompass profound temporal and somatic changes. These metaphors and analogies — these thought experiments — emerge directly from Darwin's observations and experiences on the *Beagle* voyage.

Darwin's voyage released extraordinary creativity in a previously intelligent but unambitious young man. Creativity thrives on contradiction and on asking questions that drive towards more than one solution. In his case, it also thrived on the tumult of new sense-impressions and on the constant relinquishing of those sense-impressions as he travelled on to different parts of the world. How to sustain the mass of material and to sort the evidence within it? Writing — field-notes, diaries, letters — allowed him to read himself anew and to constellate the vanished sensory evidence in other patterns. The strong alternations between enclosure on board ship, surrounded by so many men in so little space, and the extensive land journeys across wild terrain travelling only with his servant and a local guide, stimulated different forms of solitude: on board ship he was working pressed on by others but alone in thought, with only the select examples he had brought on board — the plants and beetles, bones and rocks; on his land journeys he was small in a wide landscape or struggling almost alone through difficult terrain in the midst of a profusion of impressions. These wide alternations allowed him both to hold contradictions in suspense and to glimpse, without prematurely stabilising, ideas that might make these contradictions cohere.

Though we speak of the *Beagle* as going round the world, its main task was to survey the coastline of South America and to map the seas for the safety of sailors. For long-distance travellers by ship the shipboard life may become a homeland from which to survey the changing world. The tight hierarchies of the ship mimic and exaggerate the social stratification of the home base. Comparison with an ever-receding England beyond the theatre of the ship was certainly an important intellectual resource for Darwin, but one in which England did not remain a stable point of reference; it was itself subject to re-appraisal from a distance. The British ship, the *Beagle*, moreover offered no island of stability. His stormy relations with Captain Fitzroy, the cramped conditions (sharing a cabin with 20 chronometers as well as the captain), and his constant sea-sickness drove him ashore at every opportunity. Darwin, forever escaping the boat, felt himself constantly hurried to catch up with it again at its next port of call. It was crucial that while on land he record and collect; while at sea he sorted and dreamed — and suffered.

Darwin on the five-year voyage of the *Beagle* was beset by seasickness and homesickness: seasickness, which obliterates all except the present hideous moment, and homesickness, which oppresses the imagination with flashes of past happiness and familiar scenes now utterly out of reach. Seasickness wipes out memory, homesickness excruciates it. These two ills were intercalated with ecstatic visual response to new kinds of landscape, fierce disgust and relish in encounter with indigenous peoples and their customs, and eager curiosity concerning both geological and organic life during his long land voyages of exploration and enquiry. The struggle to encompass present impressions and to conjure some understanding of the vanished past — a past that, so far as South American societies went, he could not share — took most of his imaginative energy. This young — very young — man was without any specific long-term aim on the voyage, instead he was observing with all his strength, conscious that the scenes he visited would pass by, could never be regained, and would be retained only as personal memory and record. Memory became the medium for his professional imagination, memory reinforced by inscription — in letters, in his diary, in field notes, in the revision after revision that brought all these together in the third volume of the *Journal of Researches*, which we now call *The Voyage of the Beagle*. But it was accompanied by a gathering realisation that his burgeoning theory might demand the jettisoning of human memory as a tool if a true history of the earth was to be written — and, yet further, that human memory might prove functionless in his new theory of succession and change.

The new history he was beginning to imagine stretched the resources of natural history to the limit, and beyond. It was preoccupied not only with the present variety of organic life on earth but, even more, with periods of the past when there was no human memory: not ours, not yours, not the 'others', not theirs: a world occupied by instinctual life-forms (what he succinctly calls 'knowledge without experience') and by the fluctuating movements of the earth's crust and interior. Moreover, in his determination to set his new ideas over against those of Lamarck Darwin soon came to emphasise that acquired characteristics — things learnt within the lifespan of the individual organism — cannot be handed on to progeny. Stretching your neck to reach the fruit on high branches, the well-known giraffe example, does not endow your children with longer necks. Simply, the longer-necked ones from among the brood are more likely to reach the fruit and survive.

Darwin spent much time parcelling up specimens, and describing what he had seen, but always with a sense of unrecordable excess:

> I have been very lucky with fossil bones; I have fragments of at least
> 6 distinct animals; as many of them are teeth I trust, *shattered & rolled*
> as they have been, they will be recognised. I have paid *all the attention* I
> am *capable* of to their geological site, but of course it is too long a story
> for here. [2]

That excess ('too long a story for here') is also paradoxically related to the sense of insufficiency, of meanings that recede before his eyes: he continues a little later in the same letter written to his mentor Henslow from Monte Video (Buenos Ayres): 'I purchased fragments of some enormous bones; which I was assured belonged to the former *giants*!!'[3] A world of story and belief, untapped, perhaps untappable, is caught in that remark — story that charts an alternative past, itself now gone by, perhaps in the nature of story always already gone by, 'the *former* giants'. Indeed, the mysterious relationships between past and present, through geological changes as much as migrations of peoples, are what most fascinate Darwin during his South American journeys, and what most stir his theorising imagination:

> This wonderful relationship in the same continent between the dead
> and the living, will, I do not doubt, hereafter throw more light on the
> appearance of organic beings on our earth, and their disappearance
> from it, than any other class of facts.[4]

And again:

> It is impossible to reflect on the changed state of the American conti-
> nent without the deepest astonishment. Formerly it must have
> swarmed with great monsters: now we find mere pigmies, compared
> with the antecedent, allied races.[5]

Darwin at this period of his life (1832) thought of himself as a geologist. His imagination was stirred as much by the material crust and striations of the earth, with their evidence of life millions of years ago, as it was by the short-lived beetles, flowers and creatures of the tropics, or their peoples.

2   Frederick Burkhardt and Sydney Smith (eds) (1985) *The Correspondence of Charles
     Darwin*, vol. I (Cambridge: Cambridge University Press) p. 280, 24 November 1832 to
     J.S. Henslow.
3   *Ibid.*, p. 281.
4   *Ibid.*, pp. 174–5.
5   *Ibid.*, p. 193.

Indeed, as he approached the Galapagos Islands he wrote to his friend Fox, in a passage that reads strangely in the light of what we all now know he found there for thought:

> I look forward to the Galapagos, with more interest than any other part of the voyage.— They abound with active Volcanoes & I should hope contain Tertiary strata. — I am glad to hear you have some thoughts of beginning geology. — I hope you will, there is so much larger a field for thought, than in the other branches of Nat: History.—[6]

Not a tortoise or a finch crossing his mind as yet! Instead, vanished Islands that flowed forth and hardened from a chain of Volcanoes, co-existing with early life-forms:

> —These [vanished] Islands were covered with fine trees; in the Conglomerate I found one 15 feet in circumference, perfectly silicified to the very centre. — The alternations of compact crystalline rocks (I cannot doubt subaqueous Lavas) & sedimentary beds, now upheaved, fractured & indurated from the main range of the Andes. The formation was produced at the time, when Ammonites, several Terebratulae, Gryphites, Oysters, Pectens, Mytili &c &c lived.[7]

Indeed, what Lyell called 'the autobiography of the earth' — geology — (the first citation in the OED of the word 'autobiography') gave Darwin his most prolonged encounter with a temporality which extended far beyond human memory or records. That expanded time was a necessary precondition for his thought and it was in South America that he first engaged with the meaning of rocks, volcanoes, earthquakes and shells. He found in several localities, as the paeleontological biologist George Gaylord Simpson observes in his book *Splendid Isolation*, partially petrified teeth and bones of prehistoric mammals remarkably different from present species, but remarkably akin to each other. In Uruguay Darwin purchased a skull from a farmer for 18 pence. The skull had no teeth because the farmer's sons, Darwin reported, 'knocked the teeth out with stones, and then set up the head as a mark to throw at'.

In the course of his journey he came to realise that South America had been first a part of a greater land mass that allowed creatures and life forms to spread broadcast, then had been cut off as a vast island where organic beings could discover a great variety of ecological niches without in-

6   *Ibid.*, p. 460, 9–12 August 1835.
7   *Ibid.*, pp. 461–2, to Henslow.

vasion and where a wealth of now lost species had thrived, and then had been again joined by a land bridge to North America so that dispersal and immigration dominated. South America therefore provided a superb controlled mind-experiment for interpreting the remotest past of organic life forms and their succession.

Indeed, South America provided dramatic evidence for the dependence of current life on the conditions laid down by climate and the profound upheavals of land and sea. When the earthquake struck at Concepción, Darwin happened to be lying down observing botanically in a wood on the island of Quiriquina. Suddenly the earth was in motion much as the sea — and his long suffering of seasickness made him peculiarly sensitive to the effects of that mobility: he describes it at the time as 'something like the movement of a vessel in a little cross-ripple'[8] or 'like that of skating over very thin ice; that is distinct undulations were perceptible'.[9] On reflection, his thoughts were more *about* thought and about the unsteadying of the relations between body and world:

> A bad earthquake at once destroys our oldest associations; the earth, the very emblem of solidity, has moved beneath our feet like a thin crust over a fluid.[10]

In the speculative *Notebook* he kept when he returned to England (finished 2 October 1838) he discusses the different forms of imaginative pleasure produced by the imagery of different scientific professions and characterises himself as a geologist:

> Pleasure[s?] of imagination, which correspond to those awakened during music. — connection with poetry, abundance, fertility, rustic life, virtuous happiness:— recall scraps of poetry;— former thoughts, & experienced people recall pictures & therefore imagining pleasure of imitation come into play. — the train of thoughts vary no doubt with different people, an agriculturist in whose mind the supply of food was evasive and ill-defined thought would receive pleasure from thinking of the fertility. — I a geologist, have ill-defined notion of land covered with ocean, former animals, slow force cracking surface etc. truly poetical. (V. Wordsworth about sciences being sufficiently habitual to become poetical.) the botanist might so view plants & trees.[11]

8   *Letters*, p. 330.
9   *Letters*, p. 436.
10   *Ibid.*, p. 330.
11   Gruber (1974), p. 273.

This fecund passage with its emphasis on fertility and motion suggests how essential to Darwin's thinking was pleasure, and how in South America he found the imaginative materials for inventing process. Accompanied always on his land voyages by Milton's *Paradise Lost* with its pastoral of creation, its clashing energies, he began himself to imagine on an epic scale, reaching back into cosmogony and figuring the contradictory needs of life: for profusion, for variety, and against them, for selection.

The history of multiple change and implication which came to characterise his particular evolutionary theory ('*by means of* natural selection') depended in large measure on his experiences during his three years in the extraordinary geological, organic and social diversity of South America, between 1832 and 1835. Sometimes his curiosity is ruthless. Writing to his sister Caroline he declared 'the three most interesting spectacles I have seen since leaving England — a Fuegian savage. — Tropical Vegetation & the ruins of Concepcion'.[12]

Those three images, which he here presents each as discrete 'spectacle', will by an act of profound imagination come into relation with each other as he moves towards his central idea: natural selection. In the young Darwin, as in Keats (only a little his elder), we meet the kind of creative mind and body on whom nothing is lost, and to whom everything is simultaneously available for feeling and thinking: sounds, smells, the heat of the day, beetles in ditches, cooking and companionship, the writing of earlier travellers like Humboldt, the manifold and the particular, the individual in encounter, the tribe. He has remarkably free access to his five senses.

In his repertoire of imagining he can and does use metaphor and analogy, that most frequent means of transporting and re-shaping knowledge from zone to zone. He can also — and this is rare — use direct strong connection where previously no such association has been perceived. So with these three images: Fuegian savage; tropical forest; city collapsed by earthquake. None of them is derived from the Galapagos Islands which he visited towards the end of his South American adventure. All three of them are essential precursors to the heave of imagination that he achieved not on, but after, his visit to those islands.

Most striking is what he calls here 'a Fuegian savage'. I have written extensively elsewhere in my study *Open Fields* (1996) about Darwin's contradictory responses to the Fuegian people he encountered.[13] In understanding the creative imagination that goes into the making of his major new

12   *Letters*, p. 434.
13   Gillian Beer (1996), *Open Fields: Science in Cultural Encounter* (Oxford University Press).

concept it is crucial, to my mind, to recognise that human beings were as important as Galapagos tortoises. Darwin first encountered Fuegians on that English wooden island, His Majesty's ship *The Beagle*. Because apart from its tasking of surveying and mapping the *Beagle* had a particular errand to perform: the taking home of three young Fuegian people whom Captain Fitzroy had removed from their homeland to England, two years earlier — and whom he had promised to restore. So Darwin's first encounter with 'savages' was with two young men, York Minster and Jemmy Button, and a girl of about twelve, Fuegia Basket (their soubriquets given to them on the previous voyage). They were dressed London style, they spoke in English and Jemmy Button in particular acted as Darwin's informant on flora and fauna. They were, in fact, colonised subjects apparently more or less integrated into English behaviours, and now returning trained to be missionaries and interpreters to their people. No wonder Darwin was staggered when he encountered indigenous Fuegians more or less naked under constant rain and ice, hair matted, eyes red from the fires inside their tent-like huts, and, of course, speaking in a language quite impenetrable to him, yet proving to be capable of perfect mimicry of his speech and of feats of memory quite beyond his powers. His language vacillates quite violently between disgust and fascination.

> These poor wretches were stunted in their growth, their hideous faces bedaubed with white paint, their skins filthy and greasy, their hair entangled, their voices discordant, and their gestures violent. Viewing such men, one can hardly make oneself believe that they are fellow-creatures, and inhabitants of the same world ... At night, five or six human beings, naked and scarcely protected from the wind and rain of this tempestuous climate, sleep on the wet ground coiled up like animals.[14]

Yet these are the people and the conditions from which Jemmy Button and Fuegia Basket have come and amongst whom, when Jemmy is given the opportunity to return to London a year later, he elects to remain. Darwin saw with extraordinary immediacy how dominated life and cultures are by material conditions and saw also that the categories of the so-called civilised and savage are not stable. Elsewhere in his travels in South America he learnt to de-naturalise another of the set sequences of early

---

14   Robert Fitzroy (1839) *Narrative of the Surveying Voyages of His Majesty's Ships Adventure and the Beagle, between the years 1826 and 1836, describing their Examination of the Southern Shores of South America, and the Beagle's Circumnavigation of the Globe* (London: Colburn), vol. III, p. 236 The third volume was by Darwin.

Victorian thought: the assumption that hunter-gatherers are primitive and are succeeded by settled agrarian groups. He heard how settled Indian groups had learnt to be nomadic in the face of Spanish predations. He saw the effects of slavery. This prising apart — as well as collapsing — of naturalised categories in human behaviour was crucial to his imaginative scope. In the *Origin of Species* 20 years later he purports to exclude reference to humankind. But these early and confusing experiences of human groups, while he still thought of himself as a geologist and had not yet encompassed the capacity for thought offered by natural history, helped him to dislodge the assumptions of his home society and made room for radical re-imagining of history within and beyond the human.

What books emerged immediately from these experiences? His volume, the third in Fitzroy's *Narrative of the Surveying Voyages of his Majesty's Ships Adventure and Beagle* (London, 1839), issued a year later as *Journal of Researches into the Natural History and Geology of the Countries Visited during the Voyage of HMS Beagle* and later called *Voyage of the Beagle*; *The Structure and Distribution of Coral Reefs*, *The Zoology of the Beagle* and particularly *Geological Observations on South America*, a 240 page pemmican of a book whose writing represents the turning point in his career. He writes to Sir Joseph Hooker in 1845 'I hope this summer to finish my South American Geology, then to get out a little Zoology, and *hurrah for my species work!*' His excitement has shifted from earth sciences to evolution, but his imaging of evolution draws on the deep recognition of insistent change in the structures of the earth, as much as in species and individuals. 'I a geologist, have ill-defined notion of land covered by ocean, former animals, slow force cracking surface etc. truly poetical.'[15]

Darwin travelled always with Milton's *Paradise Lost*: that poem provided both the intensity of detail and the scope of universal story to feed his imagination. He was always the foreigner, dazzled and exhausted by the strangeness of so much he was seeing, learning enough Spanish to get by, but having recourse always to English scenes and English language to measure experience. Yet, he was simultaneously learning to measure England and to understand how much vaster and more varied was the world and its peoples than England could prepare him for. Thus, by comparison, by scepticism, by re-appraisal, by immersion too, he began to construe the physical world askance the expected reading with which his culture had furnished him.

---

15   Gruber (1974), p. 273.

He was prepared to dream as well as to argue and he saw that

> a castle in the air, is as hard work ... as the closest train of geological thought. — The capability of such trains of thought makes it a dis-coverer, & therefore (independent of improving powers of invention) such castles in the air are highly advantageous.

But he remarks a curious property of such thinking as opposed to serial argument:

> The facility with which a castle in the air is interrupted & utterly forgot-ten, so as to feel a severe disappointment because train cannot be dis-covered. (In a real train of thought this does not happen. Because papers etc. etc. round one. One recalls the castle by going to the beginning of the castle) — is closely analogous to my father's positive statement that insanity is only cured by forgetfulness. — & the approach to believing a vivid castle in the air, or dreams real again explains insanity. —[16]

Darwin is here struggling to find a language for the subconscious — and also for the free play of intellection poised at a level below self-criticism; a level free of the inhibitions most associated with current and implicit cul-tural assumptions. His experience on board ship of leisure to dream, and ashore of crammed experience, made of his South American years a web of images and after-images braided with unscrutinised, uninhibited think-ing. Moreover, his South American experience started before he reached there, in the encounter with the westernised young Fuegians aboard ship. That made for the shock of his encounter with native Fuegians in which, as in dream, his shipboard companions became their own others: naked, uninterpretable, speaking in tongues and able to mimic forthwith the lan-guage of Darwin and his companions. I have written extensively elsewhere about the dreamlike nature of those encounters and his ricochets of re-sponse.[17] Not only tortoises and finches, but human beings in their native and their exotic English settings, made him know capacities for change that went beyond his own culture's present understanding.

Those processes of change required forgetting as well as memory, ex-tinction as well as continuity. In the Pampean formation he encountered mammalian remains of a plethora of extinct forms: Mastodon, Toxodon, Scelidotherium, Macrauchenia, Megatharium, Mylodon and Glyptodon.

---

16   *Ibid.*, p. 272.
17   'Four Bodies on the Beagle: Touch, Sight and Writing in a Darwin Letter,' in Beer (1996), *op cit*, pp. 13–30.

He identified the remains of a true Equus, demonstrating that species of horse had existed and become extinct before their re-introduction by the Spaniards in the sixteenth century. He heard how settled Indian tribes had responded to the predations of the Spaniards by becoming nomadic — against the ideological sequence of value then imposed in ethnography and race-theory in which the nomadic was prior and inferior to the settled. The unsteadying of apparently stable categories for describing human history, the de-naturalising of assumptions about race, as well as his experience of the instability of the earth itself , were the means of new imagining, new thought, that generated *On the Origin of Species by Means of Natural Selection* (1859).

2

Walking Backwards to the Future:
Time, Travel and Race

Benigno Trigo

A violent blow to the head removed me from the contemplative mood inspired by the sublime natural spectacle. The scene's only drawback was the ridiculous way in which it appeared, for I saw it as we walked away. The blow made me start with fright. I disobeyed the explicit instructions of the carrier, who had warned me against any sudden movement. The *carguero*[1] reproached me, and stumbling on his three feet said:

'Don't even think of it, young master.'
'Even if you kill me?'
'Not for the moment, *blanco*! But we hit a head-bender.'
'It seemed more like a neck-breaker to me.'
The *carguero* ignored my angry reprimand, pretending he had not heard it, bent over, and brought us safely from under the proverbial yoke.
Santiago Pérez, *Museo de cuadros de costumbres*

Humboldt's Cosmic Time and Temporal Scarcity

It has been argued that one of Humboldt's principal discursive strategies was 'to reduce America to landscape and marginalise its inhabitants'.[2] Consequently, his I/eye has been fixed at the crossroad of conflicting informational and experiential discourses preparing a fantasy of dominance,

---

1 *Carguero* literally means carrier. Another name for them was *sillero*, which is a reference to the chair strapped to the carrier's back into which the traveller, in turn, was fastened. According to most travel narratives, it was a profession mainly held by so-called Indians from the high plateaus, although some travellers point out that this was not always the case.

2 Mary Louise Pratt (1985) 'Scratches on the Face of the Country; or, What Mr. Barrow Saw in the Land of the Bushmen,' *Critical Inquiry*, vol. 12, p. 128.

or what has been called the strategy of the monarch that surveys all.[3] But it can be further argued that Humboldt's works also diagnose the breakdown of that monumental subject. Indeed, they stand as monuments to the consequences of gargantuan imperial conflict and overextension.

Humboldt's imperial fictions emerge as precarious attempts to map, order and extend the frontiers of empires in singular disarray. The precariousness of these frontiers is revealed by Humboldt's discourse on time and space. On the one hand, desiring an orderly world free from conflict, he set out to find temporal analogies between empires, indeed between continents and races, that seemed to his contemporaries to be terribly distant from one another.[4] On the other hand, Humboldt was seduced by the romantic rhythm of dying empires, and by the nostalgic pleasure of writing a personal narrative of his travels, even as he felt afraid of those same revolutionary rhythms, and of the resulting rupture with his own past. These fears and desires set free in his narrative the anxieties of a wandering subject a great distance away from his disintegrating country, and increasingly aware of his own mortality.

Together with these domestic and personal concerns about empire and revolution, Humboldt's works also voiced a great interest in, and a concern about, the existence and agency of an American Other. Like many of his contemporary travellers, following the romantic ideal of a union between subject and cosmos, Humboldt drew ever closer to an American subject that he helped to shape. As the gap between subject and object narrowed, Humboldt tried to keep his scientific distance with classifications and taxonomies, and felt the force of the resistant Other rebelling against those mapping practices. These struggles resulted in what González Echevarría has perceptively described as an asymptote between subject and object that further turned the expedition to the Americas into a voyage of self-discovery.[5]

Catering to the modern and commercial subject that speeds through his text saying 'forget the past and show me the future', Humboldt writes his popular *Researches, Concerning the Institutions and Monuments of the Ancient*

---

3   The eye 'commands' what falls within its gaze; the mountains 'show themselves' or 'present themselves'; the country 'opens up' before the European newcomer, as does the unclothed indigenous bodyscape. Pratt (1985) p. 124, 142n.

4   Such analogies would challenge the notions of leaders and intellectuals arguing for the inferiority of distinctly separate species, but they would also naturalize the hierarchies of the empirical mind. Stephen Jay Gould (1981) *The Mismeasure of Man* (New York: W.W. Norton and Company), pp. 30–9.

5   Roberto González Echevarría (1988) 'Redescubrimiento del mundo perdido: el Facundo de Sarmiento,' *Revista iberoamericana*, vol. 43, p. 395.

Inhabitants of America, with Descriptions and Views of the Most Striking Scenes in the Cordilleras! Researches is a collection of 'Greatest Hits', or perhaps an early form of the brochures now found at travel agencies. It elegantly adjusts itself to the modern reader's quickening pace, while preserving a reduced version of a complex network of discourses about time and space. As such it packs a more powerful punch.

In the introduction to this work, Humboldt argued for a temporal and physical continuity between the so-called old and new continents. For Humboldt, the American continent was older than had been previously thought and its landscape was as ancient as Europe's. Geological catastrophes were global, as were the migrations of species and races. Accordingly, the fables, myths and even the conception of time of the ancient inhabitants of the American continent were similar to those of the Etruscans, the Egyptians and the people of Tibet. Even if it was 'cut off from the rest of mankind' and was far from the Greek and Roman race, Humboldt argued that the American race bore a striking resemblance to the races of the 'old' continent, specifically to the race of the 'Mongol nations'.

Figure 2.1: *Carguero* Crossing the Mountain Pass of the Quindío

This temporal, spatial and racial continuity is perhaps best expressed by the plates that accompanied the book, and in particular by Humboldt's striking illustration of the *carguero* crossing the mountain pass of the Quindío (Figure 2.1).[6] At first, the gaze composing the plate seems organised by a series of opposites encompassing all time and space. With one visual gesture it fixes the past and the future, the near and the far, the high and the low, all of which have familiar meanings attached: civilisation and barbarism, life and death, city and jungle. Slightly off to one side of the central composite figure, however, stands a *carguero* whose empty chair and glance reorganise the plate's composition. On the one hand, the empty chair is a visual 'que' for Humboldt's equalising gestures. The reader knows that Humboldt has refused to ride on his *carguero*. On the other hand, the second *carguero* moves the monarch's gaze away from the first *carguero*, and looks back at him, substituting the original binomial pattern of hierarchies with a triangular composition.

The plate not only moves the gaze through space, from the first to the second *carguero*, it also suggests a temporal move between the same *carguero* at different moments in time. And indeed, both figures are identical except for one thing. The second *carguero*, now free from his burden, but keeping his position, looks up and out of the picture plane, as if acknowledging the change. Humboldt's concern with the sight of a *carguero* carrying a traveller on his back seems to make the body of the traveller magically disappear from the chair of a second *carguero*. The glance transports the disembodied presence of the traveller outside the picture plane making it occupy the same space as the monarch. The combined gaze of both the tourist and the monarch then rests at the peak of a triangular composition whose base is formed by the two *cargueros*, and by a temporal dynamic that sets the *carguero* in motion and propels the traveller higher up. The new composition suggests a firmer base for the paternal monarchy, founded on the ideal of equality among siblings, but not between parent and child. But it also modifies the hierarchy represented by the composite figure of the *carguero* front and centre, by allowing the traveller and monarch to occupy the same imaginary space. And finally, it apparently safely moves everyone a step forward, figuratively advancing everyone's best interests by promoting

6   Humboldt sketched the scene in 1801. Koch in Rome developed it into a detailed landscape and the house of Duttenhofer in Stuttgart turned into an engraving. It was then published in an *Atlas* that accompanied the more luxurious editions of *Researches*, all of which suggests the aggressively commercial bent of the book.

equality between the *carguero* and the tourist, and by opening up the privileged gaze of the monarch to the bourgeois traveller.

The complexity of the plate's composition is enhanced by an idealisation of the figure of the *carguero* in the text. Unlike the *carguero* and the traveller front and centre, the second *carguero* and the abstracted traveller exchange glances. The narrative develops this exchange into a meaningful identification.

> They talk in this country of going on a man's back (andar en carguero), as we mention going on horseback, no humiliating idea is annexed to the trade of cargueros; and the men who follow this occupation are not Indians, but mulattoes, and sometimes even whites ... When we reflect on the enormous fatigue, to which these miserable men are exposed ... when we know, that their backs are sometimes raw as those of beasts of burden, and that travellers have often the cruelty to leave them in the forests ... that they earn by a journey from Ibague to Carthago only twelve or fourteen piasters ... we are at a loss to conceive, how this employment of a carguero ... is eagerly embraced by all the robust young men, who live at the foot of the mountains. The taste for a wandering and vagabond life, the idea of a certain independence amidst forests, leads them to prefer this employment to the sedentary and monotonous labour of cities.[7]

The description is striking not only because of Humboldt's anxiety at seeing optically white men like himself doing something so degrading as carrying another human being on their backs. The description is also compelling because the motivations of the 'young men' resemble Humboldt's own motivations. The only way in which he can explain the pain they suffer and the dangers they face in their travels is by projecting on to them his own desires, his wanderlust, his need to escape the sedentary life that he so feared. Paradoxically, a still-young Humboldt sees his past beauty, strength and youth in the *carguero*. But the *carguero* represents more than Humboldt's past. He also stands for Humboldt's desire for future adventure; his desire for a virginal nature waiting to be explored and mapped. From this perspective, the plate represents the traveller, Humboldt, looking at his own idealised past and future.

---

7   Alexander Humboldt (1814) *Researches Concerning the Institutions and Monuments of the Ancient Inhabitants of America, with Descriptions and Views of the Most Striking Scenes in the Cordilleras,* translated by Helen Maria Williams (London: Longman, Hurst, Rees, Orme and Brown, J. Murray and H.Colburn), pp. 66–7.

Degeneration and the Resisting Other

Humboldt's identification with the *carguero* was not only the result of his youthful fantasies, but also the consequence of his Christian charity, and of the scientific form that religious discourse began to take around the middle of the eighteenth century. His identification with the *carguero* is the result of his anxieties about degeneration, a dystopia that became very popular by the end of the nineteenth century.

We find a repetition of this move from easy closeness to uneasy distance from an American Other in Humboldt's description of the American landscape, and in particular in his famous description of the Tequendama Falls. In keeping with his cosmological perspective, he tells the story of the creation of the Falls to illustrate the similarities between the myths of the people who had inhabited the plateau of Bogotá and the religious traditions of several nations of the old continent.[8] The description ends, however, with a view of the Tequendama from below. Abandoning his lofty perspective and putting his life at risk, Humboldt describes the view after a three-hour-long descent down a ravine. The descent dramatises Humboldt's change in perspective. Humboldt's description of the falls begins by placing the narrator on a monumental and ordered temporal plane, but it ends with a personal reflection about the end of the narrator's life.

Humboldt's work articulates domestic anxieties about degeneration. Humboldt's text, illustrations and arguments, the changes in his perspective, all suggest that the primitive landscapes and the 'extinct races' that he saw in America were too close for comfort. They suggest that Humboldt, and Europeans like him, were carried by the romantic current underlying their scientific discourse, and followed these bodies down in their fall, or in their deviation from racial perfection, anxiously suspecting their own degeneration.

But if Humboldt represents images of the colonial Other determined by such metropolitan issues, his work also stands for the remains of the agency of the Other, as it traces the *carguero*'s response to the traveller's mapping practices. Indeed, Humboldt's subtle modification of the monarchy's hierarchy can also be interpreted as an adjustment to a threat implied in the plate depicting the *carguero*. The reader will remember that the plate shows a *carguero* carrying an empty chair. In so far as that is the case, the European traveller also sees his own absence. In it, he may perhaps see his own violent ejection from the *carguero*'s chair by a figure burdened with the traveller's past and future, and weighted down by the cost of such a burden: the loss of his present Self and future prospects.

8   *Ibid.*, p. 75.

These troubling encounters with the Other, combined with Humboldt's transplanted and accumulating fears and anxieties, make inroads into his network of ordering maps. Those inroads turn the grids of his virtual maps into tears. Through these holes or *lacunae* in his work Humboldt sees his European reality deteriorate. The identification with an idealised monumental past, the simultaneous awareness of an uncertain and even collapsing future and the effect of the agency of the Other, combine to highlight for Humboldt the weakened state of the Holy Roman Empire. As much is suggested in a letter Humboldt wrote to Frederick William III (King of Prussia since 1797) upon his return from his travels, only two years before the catastrophic battle of Valmy.[9]

Mollien's Notion of Time and Degeneration

Mary Louise Pratt has convincingly shown that a 'capitalist vanguard' followed Humboldt's expedition to America. During the first three decades of the nineteenth century European travellers, mostly from England and France, visited Central and South America, and were responsible for a 'narrative of success'.[10] Pratt describes it as a 'goal-oriented' narrative of sustained progress, and heroic achievement. The travel accounts presented a confident sense of Self and an optimism about the future of the European subject. They also projected on to the landscape a 'negative aesthetic' of backwardness and neglect that legitimated European interventionism.[11] Pratt further argues that the liberal, urban, Spanish American shared the aspirations of this capitalist vanguard, but 'did not adopt their discourse wholesale as their own'.[12] She concludes that, instead, South America's educated elite turned 'to the utopian American esthetic codified by Humboldt'.[13]

Pratt is right to point out the turn back to Humboldt signified by liberal projects such as the Chorography Commission headed by Agustín Codazzi.[14] And yet, a close examination of the travel account of one of

9   Enrique Pérez Arbelaez (1959) *Alejandro Humboldt en Colombia* (Bogotá: Editorial Iqueima), p. 232. For a more detailed description and analysis of this letter, please see the longer version of this essay published in my book *Subjects of Crisis* (Hanover and London: Wesleyan University Press).

10   *Pratt* (1985), p. 148.

11   *Ibid.*, p. 149.

12   *Ibid.* p. 155.

13   *Ibid.*, p. 187.

14   In fact, the commission itself owed much to earlier voyages like Humboldt's five-year expedition to the Americas. Like Humboldt, Codazzi did not travel alone, and his texts put together pieces of knowledge from different contributors. Like Humboldt's

these economic adventurers also suggests that not all of the travel narratives produced by the 'capitalist vanguard' were optimistic. Indeed, it suggests that the vanguard's critical concept of time and race laid the groundwork for the changes in the mapping and signifying practices of the South American liberal elite.

Gaspard Théodore Mollien was a European soldier *cum* merchant. Like Humboldt, he was concerned with degeneration at home and abroad, and he associated it with heredity. Unlike Humboldt, however, Mollien emphasised biology over environment. In *Travels in the Republic of Colombia in the year of 1823*[15] he argued that miscegenation between the so-called Black, the Indian and the European races deviated the body from its primitive type, and 'weakened its constitution'.[16] Such notions made adjustments to Humboldt's comparable network of time, race and space. Despite Mollien's unpopular reception by his Spanish American readers, the adjustment would prove significant to their critical perspective on themselves, and on their country as a bodyscape and a landscape in a sustained state of crisis. If the lettered elites returned to Humboldt, as Pratt astutely claims, here I will argue that they did so only through the concept of crisis elaborated by members of the 'capitalist vanguard' like Mollien.

Mollien's travel narrative describes the numerous diseases he encountered in Colombia and his description of these diseases always emphasises a temporal component.[17] The distinct temporal nature of goitre, cretinism and degeneration also relate these diseases to the legacy of the continuing revolutionary process in Europe. The so-called degenerates that Mollien saw in America were slightly different from the deviated and 'extinct races' that Humboldt saw. For Mollien, degeneration was the legacy of a change in the perception of time and history in Europe. This change

---

expedition, the Chorography Commission extended over a number of years and drew from such varied disciplines as geography, botany, ethnography and literature. Moreover, the commission also produced physical descriptions, geographic and statistical charts, reports, descriptions of manners and customs and plates of the sights/sites seen by the travellers. For a more detailed description of the composition and the texts produced by the commission please turn to *Subjects of Crisis*.

15. The first edition of Mollien's book was published in 1824. It was translated into English that same year. A year later, however, Mollien published a second edition, augmented and reorganised. My references are to this second edition: *Voyage dans la république de Colombia en 1823*, 2 vols (Paris: Arthus Bertrand, 1825). There is no English translation of this edition. There is, however, a translation into Spanish.

16 *Ibid.*, vol. 2, p. 174.

17 Please see *Subjects* for more details about the temporal components of these diseases.

was in part the result of the meteoric rise and fall of both the revolutionary regime and the Napoleonic Empire, as well as the result of the im-
probable restoration of the French monarchy. Perhaps these cycles, revisited between 1830 and 1852, were responsible for a sense of temporal repetition and historical parody best expressed later in the century by Karl Marx
and Friedrich Nietzsche. There is no question, however, that their temporal
schemes included a pathological dimension apparently substantiated by the
travels of Europeans through their own provincial states.[18]

Patrician Optimism

Probably influenced by the optimistic thrust of the 'capitalist vanguard' as
described by Pratt, many of the American patricians that organised the different governments which emerged after the revolutionary wars saw relatively easy solutions to these temporal, spatial and racial problems. Some
even argued that the problems diagnosed by the European travellers could
be used to the American's advantage. At the end of his travel account in
1823, and expressing surprise at their unbounded optimism, Mollien
records that Colombians felt that even the nation's 'insalubrious coastal climate' worked in their favour.[19] Members of the liberal elite told a sceptical Mollien that disease acted like a protective cordon or barrier prevent-
ing a European invasion. They conjured for Mollien his own (and still per-
vasive) notion of 'race' as a biological and hereditary feature susceptible to
certain diseases. Following Mollien's understanding of the degenerate individual whose 'pathological proclivities would worsen under the influence
of a noxious environment', lettered members of the privileged social class-
es threatened Europeans with their own feared degeneration.[20]

   When applying the same concept to themselves, however, they
changed its ideological charge. They suggested that long-standing accli-
matisation had the opposite effect on their own constitution, making their
hybrid race more resistant to the diseases of the torrid zone. They sug-

18.   Europe's rural population appeared to be distant in time from the modern metropol-
      itan cities to these urbane travellers. This perceived distance led to a modernising ef-
      fort, or to an attempt at self-colonisation: a process to turn 'unruly' peasants into cit-
      izens. Daniel Pick (1989) *Faces of Degeneration; A European Disorder 1848–1918*
      (Cambridge: Cambridge University Press), p. 40.
19   Mollien (1825) vol. 2, p. 288.
20   I owe this description to E.A. Williams, quoted by Melbourne Tapper (1998) *In the
      Blood: Sickle Cell Anemia and the Politics of Race* (Philadelphia: University of
      Pennsylvania Press), p. 63.

gested that years of life in the tropics fine-tuned the 'European race' to the American environment. Wielding what Melbourne Tapper has called a 'racial formalism' that would determine the discourse on health and race by the end of the century, the liberal elite thus legitimated its authority and authorised its self-government.

Patrician liberals further manipulated the racialised scientific theories of their European counterparts to argue that the effectiveness of climate as a protective barrier depended on finding a solution to a racial problem distinct from, but related to, degeneration.[21] Drawing from the medical theories of European naturalists like Jean Baptiste Boussingault (1802–87) and Humboldt, and from travel accounts like Mollien's, they suggested that different races in their so-called 'pure state' exhibited unequal physical resistance to environmental influences. From this perspective, the 'white European race' was perceived to be at one end of a continuum of immunity that had the pure 'black' and 'indian races' at its opposite pole.[22] To be effective, and lasting, the government could not be in the delicate hands of a purely 'European race'. Conversely, a purely 'black' or 'indian' race represented a powerful challenge to a hybrid 'creole race,' which was made to occupy the middle ground of the same continuum. Many patricians began to suspect that efficient government would not permit as violent an institution as slavery to remain unreformed. Thus, not surprisingly, Mollien reported in his travel account that the answer preferred by the patricians to the threat of a stronger race was different from his own. It seemed unreasonable to Mollien that many *letrados* would not reinforce the institution of slavery as a means to keeping political control.[23] 'The new government,' Mollien complained, 'has shown itself very favourable to the slaves, and thanks to a law passed only recently, in forty years there will be none remaining in the republic'.[24] He reports that instead the new government tends to substitute the

21 Tapper (1998), pp. 69–70.

22 Boussingault cites Humboldt as saying that American Indians, due to the purity of their race, are immune to goiter. Jean Baptiste Boussingault and François Desire Roulin (1849) *Viajes científicos a los andes ecuatoriales*, translated by J. Acosta (Paris: Lasserre), p. 150. Mollien, echoing the scientific theories of his time states that neither 'indians' nor 'blacks' of pure race are prone to lymphatic illnesses. Mollien (1825), vol. 2, p. 175.

23 I borrow the term *letrado* from Angel Rama's (1984) *La ciudad letrada* (Hanover: Ediciones del Norte). It is a reference to an elite class of educated or lettered men responsible for shaping and governing the different Spanish American republics after the Revolutionary Wars

24 Mollien (1825), vol. 2, p. 194.

'black' workforce with the steam engine, while replacing optically different bodies with Europeans through an active policy of immigration.[25]

The members of the Chorography Commission wrote after this initial wave of optimism had passed. They had to confront both a belated abolition of slavery and a continuing political instability. It is not surprising, then, that they were sceptical about the panaceas imagined by their patrician elders. They became sceptical about the benefits of abolition and extended the chains of slavery well beyond the apparent disappearance of the institution. For one thing, the younger *letrados* considered that the effects of slavery were long lasting, and that they were the cause of continuing social and political stasis. Drawing from European discussions about the causal relationship between slavery and degeneration, Colombian writers like Manuel Ancízar, Santiago Pérez and Manuel Pombo imagined these figurative chains as obstacles holding the American body back from the future, as well as a force pulling its body back to the past. Similarly, futuristic sounding efforts to civilise turned into symptoms of degeneration's temporal crisis. As the Magdalena River became littered with sunken steamships, the young *letrados* criticised the earlier vision of a civilising process dependent on technological advances like the steam engine. Such easy solutions came to be seen as further obstacles standing in the way of progress.

Incorporating Temporal Anxieties

If Humboldt's morphing textual empire idealised the past as well as the future, if it projected different and changing maps on to a landscape in disarray, the Chorography Commission set out to further reorganise and remap time and space. The patrician *letrados* that studied Humboldt's maps, Boussingault's medical papers and Mollien's travel narratives developed their own landscapes and bodyscapes: fixing time, space and race according to their own needs. From Humboldt they learned to seek an idealised order within the apparent chaos. From Mollien, they interiorised a balanced, but less than ideal sense of their own racial hybridity. Following on their footsteps, the younger and more pragmatic *letrados* of the commission set out to give structure to a chaos they perceived as being normative.

The members of the Chorography Commission saw the past of the colonies as chaotic. To them, the past of Colombia involved more than

25   Mollien says that the Colombians answer his scepticism about their overtly optimistic predictions by showing him 'the constant influx from England of money, immigrants and steam engines' (Mollien, 1825, , vol. 2, pp. 287–8).

oriental-like fables and myths. It also included the Spanish Conquest (with its accompanying violent legend), the legacy of the institution of slavery (abolished during the second year of the commission's work, 1851), the Wars of Independence, and the numerous and costly civil wars roughly amounting to one every ten years between 1830 and 1860.

Moreover, temporal coordinates like the past, the present and the future lost their uniqueness to a more dynamic temporal organisation based on repetition and change. Informed by experiences, desires and anxieties both similar to, and different from, those expressed by Humboldt, the members of the Chorography Commission had a different notion of time. Time for them was neither cosmic nor was it a series of unique moments within an economy of temporal scarcity. The present was not a unique upheaval similar to the revolution that threatened Humboldt's imperial gaze, insinuated by the *carguero's* empty chair. Instead, time became for them what Daniel Pick has suggestively described as 'the experience of the pathological reproduction and transformation of revolution'.[26] Chorographic time, as it were, paradoxically blended together repetition and change. Time became an endless repetition of critical moments, even as all moments of crisis were unique and irreducible to previous models. Put in yet another way, the country was described as experiencing a degenerative condition, whose specific symptoms were dynamic and transformational, but were always dangerous and retrogressive.

Similarly, the idea of ordering space by 'fixing' its coordinates gradually gave way to that of disciplining by preserving flow or movement between spaces that now appeared to be 'too closed,' and were in need of 'opening-up'. Indeed, like the members of the 'capitalist vanguard', Codazzi and his fellow travellers saw themselves as demystifying their readers regarding the state of the national landscape. They repeated but also adjusted Humboldt's commonplaces. The geological catastrophes that Humboldt had placed in the distant past and had compared to the ancient geological history of the old continent appeared differently to the American chorographers. Using the writings of Humboldt as evidence, *letrados* like Ancízar argued not only for the comparative geological youth of the Americas, but also stressed the ever-changing shape of its landscape, while contemporaries of Ancízar like Pombo fantasised about its eventual collapse.[27] Indeed, they argued that the

---

26   Pick (1989), p. 56.
27   Manuel Ancízar (1956) *Peregrinación de Alpha por las provincias del norte de Nueva Granada 1850–51* (Bogotá: La Empresa Nacional de Publicaciones), pp. 24, 486; Manuel Pombo (1992) *De Medellín a Bogotá* (Bogotá: Talleres de Tercer Mundo Editores, Instituto Colombiano de Cultura, Colcultura), p. 143.

ordered classifications and the miniature systems used by the 'thinkers from across the oceans' were not applicable to the American landscape.[28]

In further contrast to Humboldt's cosmological perspective, and different as well from the global commercial agenda of the 'capitalist vanguard', the Chorography Commission was provincial by mandate. If Humboldt's aerial perspective provided widespread information to expanding commercial empires, and the modified perspective of the English and French entrepreneurs continued to serve a similar agenda, the Chorography Commission's provincial perspective was intended to help the emerging colonial administration enforce the law. Its commission was clearly laid out by the secretary of exterior relations of 1852 in a letter published in Codazzi's *Physical Geography*. In that letter the secretary states that in a time of increasing decentralisation, and in order to maintain a general state of order, it becomes necessary to better delineate the geographic and governmental structure of the provinces.[29] *Physical Geography* was divided into chapters describing and fixing the boundaries of the provinces of New Granada to follow this mandate. As soon as each chapter was finished, it was sent to the local authorities to improve their governing practices. Already outdated by the time of its publication as a whole, the individual chapters of *Physical Geography* had already served their political purpose.

Despite these important differences, the commission's provincial perspective and policing function incorporated Humboldt's temporal anxieties. Indeed, its texts, maps and illustrations owed much to Humboldt's representation of an uncertain future, a future threatening disorder and difference, and of a temporal economy based on the concept of future scarcity and loss best expressed by Humboldt's comments about his mortality.

A plate meant to accompany Pérez's portrait of the *carguero* entitled 'The way to Nóvita in the Mountains of Tamaná' shows the composite figure of the *carguero* moving through space, but it is also an allegory of temporal anxieties (Figure 2.2). The watercolour, like the narrative which opens this chapter, represents a precarious political balancing act. In his account, Pérez contrasts the serious and self-absorbed lofty traveller with the joking *carguero*. In the watercolour, the contrast is slightly modified to show the traveller's necessary concentration in his reading, which makes him oblivious to the danger immediately under him, a danger that makes

28   Ancízar (1956), p. 24.
29   Alexander Humboldt (1959) *Jeografía física i política de las provincias de la Nueva Granada por la Comisión Corográfica bajo la dirección de Agustín Codazzi*, 4 vols, vol. 4 (Bogotá: Publicaciones del Banco de la República), p. 292.

the *carguero* smile and tread with confidence despite the dangerous fall. What in Pérez's account is a meaningful conflict designed to underscore the tensions between eternal mapping practices and actual critical moments, is here represented as a negotiated balance solution. Conjuring the figure of a trapeze artist, the traveller balances on the *carguero*'s back, as the *carguero* balances on the log.

Figure 2.2: The Way to Nóvita in the Mountains of Tamaná

The watercolour also shows a temporal balancing act. The path just traversed is almost invisible and barely glimpsed by both the absorbed reader inside the illustration and by the viewer outside the watercolour. What lies ahead is likewise only hinted at by the invisibility of the bridge's landing and by the bend in the road, corner right. The stream and waterfall whose origins are just as invisible as their end further suggest time. Anxiety over an uncertain future, a sense of threatening disorder, an impending sensation of loss are all part of the watercolour, but unlike Humboldt's illustration, here they are front and centre. Indeed, the subject of this painting is the response to a critical moment in time: the fear surrounding the *carguero*'s very next step.

The moment is critical, its outcome may be uncertain, but the moment is certainly not unique. Pérez highlighted that lack of uniqueness, or the repetition in time of the same moment of crisis, in his account of the *carguero*. Having survived the obstacle on the road, the traveller looks back at the broad tree-trunk that hit him, and conjures the image of numerous past and future travellers paying the same toll of passage: an eternal repetition of critical moments whose outcome is indeterminate.[30] The present, then, is not a unique upheaval but a familiar moment of crisis, even if its outcome is uncertain.

Adjusting the Distance from the Other

Thus, the rejection of patrician optimism implied that the nature of the problem had not changed since Mollien, and that it was just as pressing for the members of the Chorography Commission. Indeed, Mollien's site of racial and temporal crisis was successfully interiorised, and developed, even by the European head of the Commission. A composite picture of this degenerate site emerges when Codazzi's early memoirs of his campaign during the Wars of Independence are compared with the later texts written during the Chorography expedition.

In his 1825 account, Codazzi does not associate the loss of blood with the wars against the Spaniards. Instead, he focuses on a parallel war against nature, whose blood-sucking insects and vampires multiply in sites that are always close to bodies of water.[31] Codazzi further emphasises the link between environment and loss of blood when he compares the 'diabolical'

---

30   Santiago Pérez (1973) *Museo de cuadros de costumbres*. 4 vols (Bogotá: Biblioteca Banco Popular), vol. 2, p. 152.

31   Agustín Codazzi (1973) *Memorias de Agustín Codazzi*, edited by Mario Longhena, translated by Andres Soriano Lleras and Fr. Alberto Lee López (Bogotá: Talleres Gráficos del Banco de la República), pp. 353–4.

effect of the American landscape to the loss of the body's extremities to gangrene.[32] Codazzi's early focus on blood-loss and his references to the fragmented 'European body' were strategies familiar to European travel accounts, similar in their attempts to distance themselves from a disease that was already interiorised.

Codazzi's later work, however, suggested an intensification of his fear of an internalised degeneration. There he challenged the European claim to intellectual superiority. He argued that the 'vain European race', claiming 'supposed natural privileges', would be morally and intellectually humiliated if it found itself under circumstances similar to those of the 'barbarous Indian'.[33] Moreover, he substituted his early metaphor of a foreign body loosing its blood and its members, with a national body whose members were suffering from tumours caused by lack of proper circulation in the blood.[34]

A similar repetition and amplification of the degenerate site informed Ancízar's work.[35] He too was concerned about a nation steadily headed towards depopulation, and with the speed of its degeneration. Unlike Codazzi, however, Ancízar develops the site of degeneration while grammatically implicating himself in it. Following Humboldt's and Codazzi's descent down the dangerous falls, Ancízar begins *Alpha's Pilgrimage*[36] with an ominous image of a waterfall and ends it with the closing image of a whirlpool dragging down a collective 'we'.

Ancízar's description is, in turn, an appropriate precursor to Pérez's account of the Chocó as a river fragmented (*quebrado*) by its numerous anonymous and atemporal streams, which if they blend, mix into one confused and confusing mass.

> Intricate and thick jungles stretch forever along both riverbanks. The vegetation is barely contained, and through it run a multitude of differently sized streams. Uncharted, nameless and without a history, these streams come to an end in a current where they all become jumbled.[37]

Pérez's describes a river eternally fed by uncharted and different streams that are outside time (without a history). His description suggests that Humboldt's earlier concern with temporal scarcity has been transformed

32   *Ibid.*, pp. 359–60.
33   Humboldt (1959), pp. 409, 439.
34   *Ibid.*, p. 344.
35   Ancízar (1956), pp. 48, 65, 137.
36   Ancízar used the pseudonym Alpha in *Peregrinación de Alpha* (1956).
37   The translation is mine. Pérez (1973), vol. 4, p. 175.

into a normative sense of temporal repetition and change. The river is eternally fed by disordered strands of disconnected time. As such, Perez's description repeats Codazzi's metaphor of the body politic as an organism whose blood is not lost but runs outside its proper banks. Pérez projects Codazzi's organic metaphors on to bodyscapes like the figure of the *car-guero*, a forward-moving figure in 'reflective backwardness'. But Pérez also saw the problem differently to Codazzi. Like Ancízar, he assigns a grammatical 'we' to the composite figure.[38] The chorographers increasing attention to their kinship with the object of analysis points to their concern with modes of self-government and self-definition: perhaps the two main results of the Chorography Commission.

Military Roads and Parallel Times

The crisis returned in different forms and it suggested the need for a new approach to regeneration. The cure should be different from the solutions proposed by the patrician elders; it should not be quick and violent; it should not be imposed from the outside; its form should be better adjusted to the form of the disease. As such, the cure's constitutive elements became versatility and speed, and its watchwords became movement and progress. Following Codazzi, the Chorography Commission changed the streams criss-crossing the country into metaphorical blood vessels running through a living organism. Its main recommendation was the construction of better channels of communication to facilitate, and also to regulate, the movement of different bloods through an unruly landscape that stood for a normatively diseased body.[39] In short, the Commission called for provincial roads, a less traumatic and more practical intervention than isolating and controlling 'contaminated bodies' in order to replace them with 'healthy' European specimens.

Provincial roads, or better still, military roads (*carreteras militares*), appeared side by side with police in the texts of the commission.[40] Indeed, police or *policía*, both as a set of laws and as the group of people charged with enforcing them, is a key concept to understand the function of the roads imagined and laid out by the Chorography Commission. Roads were believed necessary to restore order and direction to a population of recently freed-slaves, whose slothfulness and idleness were now seen as the result of

---

38   *Ibid.*, vol. 2, p. 149.
39   Humboldt (1959) *Jeografía*, p. 344.
40   Ancízar (1956), p. 68.

natural accident, and not as the fallout of biology and heredity as Mollien and the patrician *letrados* had suggested.[41] Provincial roads were the imagined solution to the slave's dangerous newly-gained independence.[42]

But provincial roads also suggested the ideal of self-administration, as money, resources and commerce flowed directly into local, not foreign, hands. Roads represented a self-policing ideal not only appropriate to a more federalist type of government but also to a change in the liberal political climate that had finally enforced the abolition of slavery. The new concern for the political rights of 'blacks' refocused the attention away from violent means of control to personal responsibility. A focus on duty and morality became the point of intervention of an emerging government grappling with the notion of human rights.[43] 'Blacks', Codazzi argued, could and should police themselves.[44] Provincial roads were a fitting means to this end, playing an important role in turning 'unruly blacks and Indians' into citizens, teaching them how to practise discipline on themselves.

If the long road to the internal moral and physical regeneration of the provinces called for a different sense of space and order, autonomous from the central liberal administration, it also called for a sense of time better adjusted to the nature of the problem. Clearly, more time was needed. It was now thought that long exposure to a supposedly contaminating environment had lodged degeneration deep in the American constitution, and that it would be a long time before regeneration.[45] This led Codazzi to extend the deadline for finding an effective cure, which postponed the nation's destiny and allayed the fears of its citizens.

Oblivious to the catastrophic predictions of doom-sayers, the commission instead emphasised practical movement within a parallel time frame, what Benedict Anderson has aptly called 'a complex gloss on the word meanwhile'.[46] If the river of time flowed inexorably downstream, the multiplying textual and material works of the commission opened a parenthesis, made digressions that would delay the impending end of the metaphorical river. Like the national streams in Pérez's image, the commission was consciously ahistorical. Its end was postponed over and over

---

41   *Ibid.*, p. 82.

42   Agustín Codazzi (1973) *Memorias de Agustín Codazzi*, edited by Mario Longhena, translated by Andres Soriano Lleras and Fr. Alberto Lee López (Bogotá: Talleres Gráficos del Banco de la República), pp. 333–4.

43   Pombo (1992), p. 116.

44   Humboldt (1959) *Jeografía*, p. 336.

45   *Ibid.*, p. 437.

46   Benedict Anderson (1993) *Imagined Communities* (London: Verso), p. 25.

again, even as it produced texts and maps that became obsolete soon after they were written. The government complained, withholding its funds, but Codazzi did not stop. He kept going 'without rest and without haste,' focusing on constant movement, and oblivious to deadlines.[47] The government was forced to go along until Codazzi's death, if only with recalcitrance. It was as if the Colombian nation's regeneration had become contingent on the extension of the tenure of the Chorography Commission, on the sustained dissemination of its eternally actualised reports.

Autonomous spaces and parallel times called for self-sacrifice, and Codazzi paid dearly with his own life — he famously died of malaria during his ninth, and final, expedition. Ironically, the commission's strategy for national order similarly called for creoles or *blancos* to leave the scene as soon as the road, or the law, was laid down. (By the end of the century, many of them would escape the provinces altogether, and would set off to wander through Europe instead.) In the new order, 'whites' could wander the provinces, but only as impartial and invisible observers. They knew that making room for an autonomous space, and setting in motion a parallel time would inevitably lead to mistakes (*yerros*); but, they also believed that regeneration would only come after such mistakes, after wandering off-course (*errando*). *Letrados* feared that regeneration would never come through impatient, direct and violent agency.[48]

This wandering strategy highlighted the need of an invisibility over which 'whites' had, in fact, little control. In Pérez's description of the *carguero* he emphasised the traveller's invisibility by pointing to the *carguero's* indifference to the traveller. Not only does the *carguero* erase his individuality with the suggestive epithet 'white' but he also blissfully ignores the traveller's impassioned reprimand, and holds him instead to an uncomfortable law of immobility. As we have seen, concerns over such disregard run through the travel narratives of Humboldt and Mollien. Pérez's description, however, sets him apart from the heroic stance of the 'capitalist vanguard' because it does not represent the traveller as a hero when faced with similar dismissals.[49] Neither does it occu-

---

47   Beatriz Caballero (1994) *Las siete vidas de Agustín Codazzi* (Bogotá: Carlos Valencia Editores), p. 147.

48   Pombo (1992), p. 115.

49   Mollien vividly relates an incident on his way up the Magdalena River which makes him a hero in the eyes of his boatmen. The rope that secured Mollien's canoe breaks in the middle of the rapids and the *bogas* or boatmen jump out of the boat giving it up for lost. They also abandon Mollien inside the drifting canoe, who does not know how to swim, leaving him to his own devices. 'Deafened, by the roaring of the waters and incensed by the cries of my fugitive boatmen, I leaped into the water which came

py the lofty perspective of Humboldt who wishes for an idealised realm of racial harmony. Pérez instead suggests that the *carguero's* joking dismissal of the traveller, is the appropriate response to the situation. It is a necessary allowance for the traveller's own good. Another *letrado* traveller (David Guarín) put it succinctly when he wrote: 'In the end I came to realise that not only nobody noticed me, but that I was even a threat to me'.[50]

Pérez's reference to the proverbial yoke [the *horca caudina*], however, also conveys the humiliating nature of the concession exacted on the traveller. Thus, his allegory is also critical of the necessary strategies. It suggests that the concessions are also a capitulation, that the traveller has been defeated in his struggle for authority with the *carguero*. In this way, Pérez's allegory, and by extension the works of the Chorography Commission, were not only critical of the lofty immobility of the earlier travellers, but were also critical of the self-sacrifice suggested by the practices of the present ones. Like Humboldt's representation of the empty chair, the *letrado's* frustration responded to an interiorised diseased condition, and to the agency of a resisting Other.

In sum, the sense of Self that emerged from Humboldt's work was an incongruous metropolitan subject whose perception of the Other was mediated by local discourses (historical, political, religious and naturalist). Moreover, it was a subject who returned to the metropolis with its perspective changed by Humboldt's attempts to map and order the Other. Similarly, the Self that emerged from the mapping practices of the Commission was a figure more conscious of the uncertainty of its foundational role than the founding fathers, the patrician statesmen imagined themselves to be. Ancízar best described it when he compared the Colombian creole to newborn pilgrims of uncertain future, moving like shadows through a legacy of problems and conjectures.[51]

---

up to my chin. Availing myself of an oar, which I had seized at the moment of the accident, I used it as a lever with which I lifted up the boat.' Mollien (1825), vol. 1, p. 63. See also illustrations of French travellers such as those by André Edouard's for visual examples of the traveller as hero overcoming adversity.

50 Translation is by the author. Pérez (1973), vol. 1, p. 263.

51 Ancízar (1956), p. 486.

3

National Parts and Unnatural Others: A Reflection on Patrimony at the Turn of the Nineteenth Century

Sylvia Molloy

As a child growing up in Argentina I had a particular taste for necrological notices and spent considerable time inventing lives around the spare five or six lines, boxed in small print, that *La Nación* saw fit to devote to the departed. One category of *difuntos* particularly caught my fancy, mainly because it was so rare by the 1950s, that of 'Expedicionario del Desierto'. The category comprised those who had participated in one or several of the government's concerted campaigns against Indians at the end of the nineteenth century, 'civilising' campaigns pushing the frontiers of native American 'barbarism' further and further south of Buenos Aires into the 'desert' of Patagonia, exterminating whole cultures and claiming native territories. These veterans were sometimes also called 'Héroes del Desierto', an extension of the title given to General Julio Argentino Roca, president of Argentina and the main architect of this systematic extermination.

Needless to say, very few of these *expedicionarios* were still around when I was a child, since the last military expedition led by Roca took place in 1879. That made finding one of them among *La Nación*'s terse funeral announcements all the more precious. I think I even recall one honoured with a real obit, two or more columns, under the title 'Muere el último expedicionario del desierto', although my memory may be playing tricks on me. How indeed would the newspaper have known that this was, really and truly, the last living member of one of these unsavoury campaigns that the newspapers, Argentine history, Jorge Luis Borges[1] and possibly myself as a child, unfailingly considered heroic?

---

1   '*La conquista y colonización de estos reinos – cuatro fortines temerosos de barro prendidos en la costa y vigilados por el pendiente horizonte, arco disparador de malones – fueron de tan efímera operación que un abuelo mío, en 1872, pudo comandar la última batalla de importancia contra los indios, realizando, después de la mitad del siglo diecinueve, obra conquistadora del dieciséis*' (Jorge Luis Borges, *Obras completa*, Buenos Aires: Emecé, 1974, p. 107). ['The conquest and coloni

Much has been written on these southern expeditions, especially by the veterans themselves, anxious to preserve their names for history through personal testimonials or just eager to tell a good yarn: this was the case of Lucia V. Mansilla, whose *Excursión a los indios ranqueles* — an ambivalent narrative, as much a prank, *una calaverada*, as a real expedition, whose self-deprecating tone questioned the very validity of the imperial venture — has become an Argentine classic. Yet parallel to these military campaigns there were other *expedicionarios* to the Argentine south, other desert heroes who conceived their mission in much the same patriotic terms as did the military, and whose weapons, no less destructive than firearms, were also at the nation's service. Self-styled scientists, archaeologists, botanists, geographers, anthropologists, and usually a combination of all these categories, they too travelled to the south, especially to Patagonia, less intent on pushing the Indian away from the city than on bringing him back there, albeit in pieces, as an object of study, an item for a collection. One of these also stands out in my memory from childhood days (and this is the last time I shall resort to personal history, at least overtly) by the way in which he was usually referred to, never by his full name, Francisco P. Moreno, but by the function assigned to him, *Perito Moreno*, 'Expert Moreno'. An unusual appellation, to say the least, it has also become the name of a geographical site occupying a place of privilege in the Argentine national imaginary: I refer to the monumental Perito Moreno Glacier in southern Patagonia, one of the last glaciers in the world that continues to expand, and a symbol of sublime, ever-growing 'Argentine-ness' in the most remote confines of the nation.

Institutionally speaking, anthropology, archaeology and ethnology were latecomers amongst the disciplines that founded the nation in the nineteenth century. History was, of course, the most important discipline, since it urgently needed to be rewritten. Secession from Spain immediately triggered debates on historiography; innumerable historical 'collections', 'pantheons', 'archives', 'galleries' — these are all words used in actual titles — were published in the mid-nineteenth century throughout Latin America narrating the feats of the new Latin American heroes. Literature itself was perceived as a subordinate discipline, historical novels, patriotic poetry and

---

sation of these domains – a handful of fear-ridden mud forts clinging to the coast and watching the curved horizon, a bow that shoots forth Indian raids – was so indecisive that, in 1872, one of my grandfathers was to command the last major battle against the Indians, bringing the sixteenth-century conquest to a conclusion only after the middle of the nineteenth century'. (Jorge Luis Borges, *Evaristo Carriego*, translated by Norman Thoams Di Giovanni, New York: Dutton, 1984, 42).

political manifestos being, for a long time, its preferred vehicles. Parallel to these historiographical efforts, Latin American nations, heeding the call in Bello's *Silva* to catalogue the local, strove to construct their own, national classifications, dictionaries and taxonomies, describing nature, mapping territories, collecting data *pro patria*. It is not my purpose here to discuss the formation of disciplines and their specific discourses in Latin America, merely to stress the tentative, *cumulative* aspect of all these institutional, or protoinstitutional ventures, their *collective* nature, in all senses of the word. Historical 'pantheons,' no less than glossaries or botanical lexicons, are collections, deliberate 'gatherings' (of military heroes, of botanical specimens) designed to signify both the nation's symbolic wealth and — in the case of collections constructed by the incipient natural sciences — the nation's potential for very real material wealth. It is within this frame, that of the national collection, that I wish to consider Francisco Moreno, whom I too shall call, for the sake of expediency and perhaps because I cannot do otherwise, Perito Moreno.

Among the many travelogues and reports he wrote, I shall focus on one text, *Viaje a la Patagonia Austral*, the most exhaustive summary of his many voyages south. Out of print for many years, the book was reissued in 1997. In an unsubtle effort to give new currency to the author and ensure a reading of the book along specific ideological lines, the publisher took inordinate care to provide meaningful biographical details and an enthusiastic critical evaluation. Thus, the fly-leaf informs the reader that Moreno was a precocious explorer (he was only 21 when he undertook his first trip south); that he suffered great hardship during his travels due to 'the precarious conditions of the period and the lack of official support for his enterprise'; that he carried out his travels 'spurred by his investigative spirit and by his urge to demonstrate the possibility of populating and civilising the immense Patagonian territories, still unknown and uninhabited, yet already awakening the curiosity and greed of more than one foreign nation'; that in 1886 he donated his collections to the nation to create a Museum of Natural History and was appointed its director for life; that in 1897 he was named official Argentine expert ('perito') in the boundary disputes between Argentina and Chile; that in 1912 he travelled to Patagonia for the last time to accompany Theodore Roosevelt; that, finally, he was 'an erudite expert in the Indian question, (disapproving of the cruel solution that was then applied), a scholar well respected in the most important European universities' and that 'a profound personal unselfishness and an exemplary patriotism inspired his untiring travels'. Although it would be tempting to go into the strategic timeliness of this republication in 1997, at a time when Patagonia, once again, is perceived as the object of 'the curiosity and greed of more than one foreign nation', I will stay with the construction of the scientist as national hero and take a closer look at travel and collecting as patriotic duties, and the collection as the principle of national patrimony.

Science and *patria* are closely linked in Moreno's book from the start. Eager to cultivate the figure of the self-effacing scientist, Moreno states that he is well aware of his shortcomings as narrator when compared with his predecessors, Humboldt and Darwin. Yet he brings his readers something those master narrators forcibly lacked: Argentine patriotism, conceived both as a national feeling for territory and a national duty to lay claim on and exploit that territory in the name of national progress. Thus Moreno writes:

> *Sólo aspiro a que con esta narración mis compatriotas puedan formarse una idea de lo que encierra esta gran porción de la patria, siempre denigrada por los que se contentan con mirarla mentalmente desde las bibliotecas.*

> [I only hope that, with this narrative, my compatriots may form an idea of what this great portion of our country (la patria) contains, a portion usually disparaged by those content to just contemplate it mentally, in libraries].[2]

He adds, at the end of the preface:

> *[D]esearía que [esta lectura] contribuyera a que algunos de mis compatriotas visiten las regiones que describo [...] haciendo votos para que los colores patrios que dejé solitarios en el punto más lejano que alcancé durante mi viaje, sean llevados más adelante por otros argentinos, en provecho de la patria y de la ciencia.*

> [[I] would hope that [this book] inspire some of my compatriots to visit the regions I describe [...] and that the national colours [*los colores patrios*] I left in solitude in the farthest point I reached during my travels will be carried forward even further by other Argentines, in benefit of the country [*patria*] and of science.][3]

Like many autobiographers intent on showing the origins of their vocation in the most insignificant childhood pastimes, Moreno dates his bent for collecting back to his early years. As an adolescent, together with his broth-

2   Francisco Moreno (1997) *Viaje a la Patagonia Austral* (Buenos Aires: Elefante Blanco), p. 6.

3   *Ibid.*, p. 7.

er and a cousin, Eduardo Holmberg, the future novelist who would also attain renown as a scientist, Moreno started 'collections' — of pebbles, of insects, of stray animal bones, of Indian artefacts — and formed a 'museum' that was much admired by adults. The moment of revelation came at the age of 20 with a singular gift:

> *En 1872, el envío de algunos objetos considerados de importancia por personas competentes, hecho por un amigo residente en Carmen de Patagones, me decidió a llevar a cabo mi primer viaje a la Patagonia.*

> [In 1872 a shipment containing certain objects, which experts considered important, sent by a friend living in Carmen de Patagones, convinced me to undertake my first trip to Patagonia.][4]

Interestingly, Moreno does not specify the nature of these mysteriously important objects that trigger his first scientific and patriotic voyage. One can guess what they were, however, from the description, in the very next sentence, of what he calls the 'abundante cosecha' [abundant harvest] that resulted from his first trip.

> *Corto fue el viaje, pero provechoso. Los paraderos y cementerios cuya existencia había revelado Strobel, me suministraron cráneos y objetos de piedra en número suficiente para poder formarme una ideal del interés que ofrecía el estudio del indígena patagónico.*

> [The trip was short but profitable. The settlements(?) and cemeteries whose existence had been revealed to me by Strobel furnished me with skulls and stone objects in sufficient numbers to allow me to appreciate the interest of studying the Patagonian Indian.][5]

Moreno, the 20-year-old patriotic scientist, collects artefacts and skulls, stones and bones.

Yet Moreno is dissatisfied with only collecting relics:

> *No bastaba estudiar las generaciones extinguidas que el tiempo había sepultado en el litoral marítimo patagónico; era necesario compararlas con las tribus que las sucedieron en la posesión del teritorio, y al efecto debía visitarlas en persona. Vivir con los indígenas en sus mismos reales y recoger allí los datos buscados vale mucho*

4   *Ibid.*, p. 10.
5   *Ibid.*, p. 11.

*más que leer todas las relaciones de los cronistas, que generalmente no son abun-
dantes en la verdad de lo que cuentan.*

[It was not enough to study defunct generations that time had buried
along the Patagonian seabord. It was necessary to compare them with
the tribes that succeeded them in the possession of these territories,
and in order to do so I had to visit them personally. To live with the
Indians in their own dwellings, gathering the desired data on location,
is much more valuable than to read the narratives of the *cronistas* which
do not abound in truthfulness with regard to their subject matter.][6]

This double vocation, as archaeologist and anthropologist, results in a remarkably complicated double gaze, a gaze simultaneously taking place, one
might say, in two distinct times. On the one hand, Moreno is gathering (or
harvesting, to use his favourite word) data on living Indians — albeit
'sometidos a la autoridad nacional' [subjected to national authority],[7] that
is, severely deprived of agency — he is taking precise note of their customs (in much more detail than, say, Mansilla), minutely observing their rit-
uals, even going so far as to praise their good qualities:

> ... *en los centros civilizados generalmente no se conocen (o no se quieren admitir)
> los instintos generosos del indio [...] Su mayor deseo es aprender todo lo que, com-
> patible con su carácter, pueda enseñarle el europeo, y si con su familia llega a con-
> seguir algunas comodidades, no vuelve jamás a su vida nómada.*

> [... civilised centres are generally unaware (or do not want to be made
> aware) of the Indian's generous instincts [...] His greatest desire is to
> learn everything the European will teach him which is compatible with
> his character. If he and his family secure a certain degree of comfort,
> he will never return to his nomadic life.]

Moreno sees the Indian as a potential Argentine subject, a lesser subject, it
is true, since he belongs to 'las especies más degradadas e inferiors' [the
most degraded, inferior species],[8] but a subject no less, whose mode of life
is worthy of study, and whom he endows with limited agency and subaltern futurity: 'El día que el tehuelche, así como las demás tribus de la
pampa, conozcan nuestra civilización antes que nuestros vicios y sean
tratados como nuestros semejantes, los tendremos trabajando en las es-

6   *Ibid.*, p. 15.
7   *Ibid.*, p. 16.
8   *Ibid.*, p. 12.

tancias del Gallegos' [the day that the tehuelches, as other tribes from the pampas, learn our civilised ways, not our defects, and are treated as our equals, we will have them working on the ranches along River Gallegos].[9] On the other hand, however, Moreno sees the Indian as an object of study, preferably in parts, as a specimen for a collection which will earn him national prestige and heroic civic stature. While the first attitude has him looking at the Indian with curiosity, as a subject worthy of ongoing study in a living context, the second has him looking at the Indian *as* curiosity, that is, as an inert, isolated item worthy of study, but above all, worthy of being exhibited. The slippage between the two may be noted in Moreno's casual syntax: 'Las dos visitas al río Negro me dieron por cosecha ochenta antiguos cráneos de indígenas, más de quinientas puntas de flechas trabajadas en piedras, muchos otros objetos y algunos cráneos y utensilios actuales' [My two visits to Río Negro allowed me to harvest eighty ancient Indian skulls, more than five hundred arrowheads carved from stone, many other objects, and a few recent skulls and artifacts'.[10] Nowhere is that slippage more striking (and more poignant) than in the chapter adequately titled 'Restos Humanos' [Human Remains], where Moreno recounts the episode of a good Indian friend he has met on a previous trip:

> *Cerca de la comisaría nacional está situado el cementerio de la colonia y en él había sido inhumado mi amigo Sam Slick, buen tehuelche, hijo del cacique Casimiro Biguá. Conocí a ese indio en mi viaje anterior a Santa Cruz [...] Nuestra llegada [...] fue un motivo de gozo para el buen Sam, por los regalos y los ponches con que lo obsequiábamos [...] Su contento rayaba en entusiasmo cuando lo embarcábamos de vez en cuando el el bote, le dejábamos manejar el timón, y escuchar el tambor y el pífano del bergantín.*

> *Consintió en que hiciéramos su fotografía, pero de ninguna manera quiso que midiera su cuerpo y sobre todo su cabeza. No sé por qué rara preocupación hacía esto, pues más tarde, al volver a encontrarlo en Patagones, aun cuando continuamos siendo amigos, no me permitió acercarme a él mientras permanecía borracho, y un año después, cuando llegué a ese punto para emprender viaje a Nahuel Huapi, le propuse que me acompañara y rehusó diciendo que yo quería su cabeza. Su destino era ése. Días después de mi partida se dirigió a Chubut y allí fue muerto alevosamente por otros dos indios, en una noche de orgía. A mi llegada supe su desgracia, averigüé*

9   *Ibid.*, p. 469.
10   *Ibid.*, p. 16.

el paraje en que había sido inhumado y en una noche de luna exhumé su cadáver, cuyo esqueleto se conserva en el Museo Antropológico de Buenos Aires; sacrilegio cometido en provecho del estudio osteológico de los Tehuelches.

[Near the police station was the cemetery of the settlement and in it had been buried Sam Slick, the good *tehuelche*, son of chief Casimiro Biguá. I knew that Indian from a previous trip to Santa Cruz [...] Our arrival then had been an occasion of joy for good Sam because of the presents and liquor we gave him [...] His happiness turned to sheer excitement when once in while we took him with us out to sea, letting him hold the rudder while listening to the drums and the fife on the brigantine.

He agreed to let us take his photograph but in no way let us measure his body, especially, his head. I don't know what strange preoccupation led to this refusal. Later on, when I again ran into him in Patagones, in spite of our continued friendship, he would not let me near him when he was drunk. A year later, when I again returned to that spot to undertake a journey to Nahuel Huapi, I asked him to come with me and he refused, saying that I was after his head. That was his destiny. Days after my departure he left for Chubut where he was treacherously killed by two other Indians during a night of orgy. When I returned I was told of this terrible outcome, found out where he was buried and, on a moonlit night, dug up his body. His skeleton is preserved in the Anthropological Museum of Buenos Aires. This was a sacrilege committed to advance the osteological study of the Tehuelches.]

Today a friend, tomorrow an exhibit. Sam Slick, the sunny, childlike, Indian who was allowed to play sailor to the amusement of the white man (Moreno's narrative casually abounds in what Borges calls pathetic circumstantial details, *the hallmark of efficient storytelling*), Sam Slick becomes part of the collection on which Moreno's museum is founded. Moreno adds that he similarly exhumed the bodies of *Sapo*, an Indian chief, and his wife, both of whom had been buried a few years before. These additional exhumation allows him to give a detailed description of Indian burial rites, including the fact that the woman had been buried with her pet dog, a fact leading to the statement 'Hay algo de poético en lo que motiva el entierro de los perros junto con los restos de los que fueron sus dueños'. [There is something poetic in this custom of burying dogs together with the remains of their erstwhile owners.] 'Con estos *objetos* y los anteriores — he concludes — quedé satisfecho sobre este punto importante de mi viaje.' [With these *objects* and the ones mentioned previously I was satisfied on this important aspect of my trip].[11]

Moreno's collection at one point comprised, as Jens Andermann notes, not just good dead Indians but 'a two-story high glass case with dozens of Indian skeletons, several of which, Moreno boasted, were of 'renowned Indian chiefs slain during the military campaigns'.[12] Additionally, to give another example, this time at an institutional level, of the slippage between live human beings and lifeless exhibits, so well illustrated by Moreno's double gaze, the collection comprised not only dead specimens. Indian chiefs and their families were taken to Buenos Aires as captives and then 'employed' by the museum in menial tasks. These Indians too were exhibits *and, at the same time*, living sources of information: 'les femmes devaient enrichir les collections ethnographiques par leur travaux de tissage en même temps qu'on pouvait étudier leurs moeurs' [women would enrich the ethnographic collections with their weaving; at the same time, one could study their customs].[13] When these captive Indians died, mostly of consumption, their skeletons were entered into the collection they had hitherto tended to.[14] Today a subaltern, tomorrow an exhibit.

How does the collection harvested by the traveller become the national public exhibit curated by the expert? Moreno explains: 'Fruto de mis tareas ha sido la colección que he formado y que he tenido la honra de donar a mi patria para fundar el Museo Antropológico y Arqueológico de Buenos Aires, del cual soy director y a cuyo desarrollo destinaré todos los años de mi vida.' [The result of my efforts is the collection I have created, which I have had the honour of donating to my country [*patria*] with the purpose of creating the Anthropological and Archaeological Museum of Buenos Aires, of which I am director, and to whose development I will devote all the years of my life.][15] In truth, the relationship between these activities, less casual than appears at first glance, amounted to a very real material and symbolic transaction. Moreno donated the collection on which the museum would be based *in exchange for* the title of director for life, thus assuring himself a position of considerable prestige in the Argentine establishment. In relinquishing one form of control to gain another, he is exchanging the private collector's

11   *Ibid.*, p. 106 (my emphasis).
12   Jens Andermann (1998) 'Evidencias y ensueños: el gabinete del Dr. Moreno,' in *Filología*, vol. 31, nos. 1–2, pp. 57–66.
13   *Ibid.*, p. 10.
14   *Ibid.*, p. 5.
15   Moreno (1997), p. 13.

'phantasmagoria of the interior' (in Walter Benjamin's phrase) — or the scientist phantasmagoria of the private research space — for the curator's phantasmagoria of the national, a phantasmagoria he is in charge of offering up, didactically, to the public in a space which is now communal: a 'grandioso templo' [grandiose temple],[16] as Moreno calls the space of the museum. Additionally, the Argentine government rewards him with thousands of hectares of Patagonian land in the choice area of Nahuel Huapi, a valuable remuneration that he will, in turn, freely give back to his country for the creation of a National Park.

If national glory means much to the *perito*, material possession does not. Among Moreno's papers, found after his death, was the following statement:

¡Tengo sesenta y seis años y ni un centavo! [...] Yo que he dado mil ochocientas leguas a mi patria y el Parque Nacional, donde los hombres de mañana, reposando, adquieran nuevas fuerzas para servirla, no dejo a mis hijos ni un metro de tierra donde sepultar mis cenizas! Yo que he obtenido mil ochocientas leguas que se nos disputaban y que nadie en aquel tiempo pudo defender sino yo, y colocarlas bajo la soberanía argentina, no tengo donde se puedan guardar mis cenizas: una cajita de veinta centímetros por lado. Cenizas, que si ocupan tan poco espacio, esparcidas, acaso, cubrirían todo lo que obtuve para mi patria, en una capa tenuísima sí, pero visible para los ojos agradecidos.

[I am sixty-six years old and am penniless. I who have given eighteen thousand leagues to my country (*patria*) and the National Park, where men of the future may relax in order to gain new strength to serve it, I do not leave my children so much as a square metre of soil where my ashes might be buried! I who have won eighteen thousand leagues that were in dispute and that no one at the time could defend as well as myself, placing them under Argentine sovereignty, have nowhere to place my ashes, a mere twenty-centimetre square box. These ashes, occupying so little space, would perhaps cover, were they to be scattered, all the land won for my country, cover it with a tenuous film, to be sure, but one visible to the grateful eye.][17]

16   Moreno (1997), p. 13.
17   Eduardo V. Moreno (1997) *Reminiscencias del Perito Moreno* (Buenos Aires: Editorial Elefante Blanco), p. 12 (orginal edition Buenos Aires: Imprenta del Estado, 1897).

The eighteen hundred leagues he won for his country from Chile, the National Park his country rewarded him with, and his own ashes turned into 'territory' are one and the same: possessions, all, which he bequeaths to his *patria*.

Ciencia, in Moreno's design, ensures the construction of *Patria*. Thus Moreno tells us that his museum 'contendrá algún día la historia de los primeros pobladores de nuestro suelo, consignada en sus obras, asistida por sus mortales despojos. Allí sus descendientes podrán estudiar sus progresos.' [one day will contain the history of the first inhabitants of our land, conveyed by their crafts and complemented by their mortal remains. Here their descendants will be able to study their progress.'[18] The loose usage of the term *descendants* on the part of the *perito* Moreno — for surely the expert knows that, given the 'cruel solution' advocated by the government, extinction of the Tehuelches is a matter of years and no biological descendants will be around to gaze on the bits and pieces that remains of their elders — marks the passage from *Ciencia* to *Patria*. In this spirit, it is interesting to note that after the 1910 centenary the museum's anatomical collection is renamed 'Panteón de los Héroes Autóctonos' [Pantheon of Autochtonous Heroes].[19] Today an Indian, tomorrow a national ancestor. This pantheon remained intact until the 1930s, when the actual bones were removed from display and replaced with wax masks and busts of Indians warriors. In 1973, the bone collections were finally returned to Indian groups demanding their proper burial.

The pieces assembled by Moreno in the course of his several expeditions to Patagonia are, it could be argued, 'natural' parts that the 'expertise' of the collector/curator then turns into 'national' parts, to be incorporated into a nation pedagogically displayed in the museum. The scramble for an autochthonous and unsullied American past (Moreno repeatedly indicates his preference for 'pure' Indians over *mestizos*, not entirely for 'scientific' reasons) is in part a reaction against ragtag 'foreign' immigration but it is also a way of laying claim to a territory in dispute. Moreno, as he goes about collecting, also stakes out the limits of the nation, those other natural parts or partings to which he gives Argentine names. As he explores Western Patagonia, where the boundary with Chile is a subject of bitter contestation, he claims landmarks as he had claimed skulls, in name of the *Patria* and with the same quasi religious zeal. 'Llamo a este cerro "Monte Félix Frías!" en honor de mi venerable amigo el esclarecido patri-

18   Moreno (1997), p. 13.
19   Andermann (1998), p. 10.

ota que defiende con tanto ardor la causa de los argentinos contra las temerarias pretensiones chilenas' [I hereby name this hill 'Mount Félix Frías' in honour of my venerable friend and illustrious patriot who so ardently defends the cause of Argentines against the brazen Chilean pretensions] he writes.[20] Or: 'Esta montaña se llamará en adelante cerro de Mayo' [This mountain will be called from now on Mount May];[21] or: 'a nuestra derecha la arqueada falda de los montes que he llamado Buenos Aires, al pie del ramal del lago que precede a los inaccesibles Andes y al norte del pintoresco monte Avellaneda, que nombro así en honor del presidente de la República' [to our right the curving mountainside I have called Buenos Aires, below it the section of the lake at the foot of the inaccessible Andes, and to the north picturesque Mount Avellaneda, which I have named thus in honour of the president of the Republic'.[22] In a remarkable passage, nature itself is pressed into service, supporting Argentina's 'natural' right to the landscape Moreno busily baptises:

A la tarde emprendemos el regreso, después de dejar como signo de nuestro paso, clavada sobre un enorme fragmento de roca testigo mudo de la poderosa erosión de los hielos, y rodeada por verdes helechos y rojas fucsias, la bandera patria que nos ha acompañado durante toda la expedición y cuyos colores copian ahora la alfombra blanca de nieve recién caída y el celeste del hielo eterno que cubre desde la cumbre el inaccesible pico Mayo.

[In the afternoon we start back, after leaving behind us, as a sign of our presence, set in an enormous boulder which mutely attests to the powers of erosion of ice, surrounded by green ferns and red fuchsias, the national flag which has accompanied us throughout the expedition and whose colours are now being copied by the white carpet of newly fallen snow and the blue tones of the perennial ice that covers the summit of the inaccessible Mount Mayo.][23]

Blue and white, of course, are the Argentine national colours: this is a case not of nature imitating art but of nature imitating nation. After a lengthy description of the southernmost Andes Moreno concludes: 'Estos son los límites que la naturaleza ha trazado entre los dos países. Las pretensiones chilenas no deben ir más allá de ellos y nosotros los argentinos no debemos tampoco consentirlo.' [These are the boundaries that nature has drawn be-

20   Moreno (1997), p. 441.
21   *Ibid.*, p. 444.
22   *Ibid.*, p. 447.
23   *Ibid.*, p. 447–8.

tween the two countries. Chilean pretensions should not cross them nor should we Argentines allow it].[24] As he says in his conclusion, the Patagonian regions he has explored and named have been, through his intercession — and I give the term intercession its strongest meaning — 'revelados a la geografía de la patria' [revealed to the geography of the nation].[25]

So much for nature in national revelations and the national policies that result from them. But what if the traveller, intent on discovery, comes upon a specimen that seemingly exceeds national frontiers? Let me turn to a text by sociologist, political scientist and writer Carlos Octavio Bunge published in 1908, in a collection of stories titled *Viaje a través de la estirpe y otras narraciones*, [Travels through the Species and Other Stories]. The allusion to positivistic science in the title is further confirmed by the fact that Charles Darwin himself appears as a character in one of the stories. The piece I am interested in is titled *La sirena* and narrates, once more, a trip to Patagonia. However, this time the traveller is quite undistiguished: not a scientist, not a particularly patriotic subject, not even a member of the aristocracy, but a businessman of Italian descent on vacation, looking to rest after demanding commercial ventures and the general turbulence of city life. His first trip south takes him to a seaside resort where, walking one night on the beach, after having gambled heavily and lost, he sees what he takes to be a siren. The following night he swims out to find her, nearly drowns and is rescued by what turns out indeed to be a siren, albeit one that hardly meets his expectations:

¿Era este monstruo, con su largo apéndice natatorio, con su coriácea piel de delfín, con su aspecto fiero y silvestre, el bello ideal de Sirena que forjara la fantasía humana y soñase yo en sueño de amor?

[Was this monster, with its long appendage for swimming, its leathery dolphin skin, its ugly wild looks, the beautiful ideal of a Siren that human fantasy has created and that I dream of in my love dreams?].[26]

His erotic delusions shattered, the protagonist, hastily turning ethnographer, proceeds to interview his 'froglike'[27] subject, is told by her that sirens avoid humans because 'nos horripila la idea de que algún día puedan pescarnos y exhibirnos vivas en sus jardines zoológicos, o bien en sus museos, disecadas y embalsamadas' [we are horrified by the idea that some

24   *Ibid.*, p. 454.
25   *Ibid.*, p. 477.
26   *Ibid.*, p. 102.
27   *Ibid.*, p. 101.

day you might capture us and exhibit us alive in your zoological gardens, or else in your museums, desiccated and embalmed].[28] The interview is carried out in Spanish because, he is told, sirens were taught Spanish by the first *conquistadores* in exchange for food. He also learns that sirens belong to a very ancient species, much older than humankind but since the appearance of man, 'nuestra raza viene decayendo y degenerando. Tal vez se extinga muy pronto' [our race is in the process of decaying and degenerating. It may extinguish itself very soon].[29] Once the interview is finished, the siren swims away and the narrator returns to his commercial preoccupations, not giving the siren further thought 'pues que no soy curioso ni naturalista' [since I am neither a naturalist nor a curious man].[30]

This is not the end of the siren. The story goes on to narrate a second voyage, by sea on an English vessel, this time to Patagonia, and more precisely to the Malvinas/Falkland Islands, then (as now) an object of dispute with Britain. The siren is sighted once more, this time captured, and, once on board, the struggle for national sovereignty begins between Robbio, the Argentine narrator, and the other passenger, an Englishman called 'Mister Phillips', also a businessman. A farcical dialogue ensues, each businessman wishing to take the siren back to his country in order to exhibit her, making jokes about the need for a Solomonic judgment, till the Argentine narrator explodes: 'No, Mister Phillips. Es toda ella mía, y para mí es cuestión de patriotismo llevarla entera a que la estudien los técnicos de mi patria y atraiga allí las miradas del mundo todo.' [No, Mr. Phillips. She's all mine, and for me it's a patriotic duty to take her whole so that she may be studied by my country's scientists and may attract the gaze of the whole world'.[31] National belonging, or the appearance of such, finally decides the argument. Both adversaries agree that if the siren, when she comes to, responds to questions in good Spanish [en buen castellano] she belongs to Argentina; if not she may go to England. And of course, the siren does. The third and last section of the story is devoted to the narrator's quandary when arriving in Buenos Aires, his mixed feelings towards the siren (he hides her in his *garçonnière* as if she were his lover and at the same time searches for an institutional space where she may be studied) and, finally, opts for the easy way out: the siren, who has been taken to a secret cage in the zoo, asks to be released back into the sea so that men may continue to dream of her and the myth may live on.

28   *Ibid.*, p. 104.
29   *Ibid.*, p. 105.
30   *Ibid.*, p. 108.
31   *Ibid.*, p. 119.

I have gone in detail into a story its author surely considered a frivolous *divertissement*, built on the model of the charming escapade narrated during a post-prandial chat, because it adds several layers to the reflection on national patrimony, scientific collections and living exhibits that I have engaged in. The interview with the siren reveals significant points of contact between her and the subjects/objects of *Perito* Moreno's collection. Like sirens, Indians were ancient inhabitants of a region and members of a race weakened by contact with (white) men and about to be extinguished, and they too had learned Spanish on contact with the Spaniards. But the siren, unlike the Indian, is not 'como nuestro semejante' [like a fellow being], she cannot be taught to work in ranches, sweep museum floors, or weave additional specimens for the museum's collection. She is, unquestionably, excessive, a monster. And it is as such — as a froglike, fishy-smelling, grinning, hybrid monster, and not as the legendary icon of seduction — that, in Bunge's story, she *unnaturally* signifies the nation, qualifying as object for scientific, patriotic study by 'los técnicos de mi patria' and for display before the rest of the world, 'las miradas del mundo todo.'

I do not know if Bunge, so intent on elaborating national cultural discourse and laying down national law in other areas, had any sense of the limits his narrative so perversely brushed against, any sense of the destabilising thrust his recourse to unnaturalness effected on patriotic travelling and patriotic accumulation such as Moreno's. He may have had an inkling in that he made his traveller not a scientific expert, not a hero, not a collector but a businessman, clever enough to 'win' the siren in the argument, but not wise enough, or motivated enough, to keep her and exploit her, *pro patria* or for his own benefit. Or maybe wiser than we think. In a period where the natural/national is roped in and put on display, Bunge's character, stealing into the zoo in the dark of night to free the siren and throw her back into the sea, is performing a salutary act of critical inquiry well ahead of his time: he is liberating the national, showing it up as an unnatural defying classification, and — ruefully, humorously — allowing it to drift away.

4

On the Transition from Realism to the Fantastic in the Argentine Literature of the 1870s: Holmberg and the Córdoba Six

Eduardo L. Ortiz

I. Prolegomena

Sarmiento's Institutions

Professor Germán (Hermann) Burmeister was a world-famous German naturalist who had been officially associated with Argentina from the 1860s. In the early 1870s he was invited by President Sarmiento's government to select and recruit in Europe a group of German-speaking scientists to help in the organisation of a new science research institute. A few years later this institute, adjacent to the old university of Córdoba, became the Argentine National Academy of Sciences. After an auspicious start, frictions developed between Professor Burmeister and the imported German scientists. Burmeister wished to control scientific life in Córdoba from his position in Buenos Aires, where he was director of the Natural History Museum.

As a result of disagreements with Burmeister, in early 1874 the six leading scientists engaged abroad were separated from their chairs and left Córdoba, and some returned to Germany immediately. Young scientists in Buenos Aires made common cause with those resigning. The event was seen, by some, as a parallel to the famous resignation, in 1837, of seven professors in Göttingen, the so-called *Göttinger Sieben*, on account of restrictions imposed on their academic freedom.

The Córdoba affair was widely covered by the press in Córdoba and Buenos Aires, as well as in other main cities of Argentina. It even had an impact on prestigious German and other foreign scientific journals. Sarmiento was put in a difficult position, as he relied on Burmeister for scientific advice. Politicians hostile to Sarmiento took advantage of the situation to criticise his handling of the situation, which let Córdoba down. These critics were not alone, even some of Sarmiento's intellectual friends and political supporters were shattered by the obstinacy of the national government supporting Burmeister, and letting the scientist go.

Students in Córdoba, some of whom later became leading figures in the country, expressed their support for the German teachers, whom they perceived as opening new nationally useful professional perspectives for them. This attitude was a break with tradition and anticipated further expressions by Córdoba's university students in relation to educational reform. Some forty years later, the 1918 Argentine university reform movement, that started at the old university of Córdoba, had a profound impact on Argentina's university system and political life, and an immediate resonance on several countries of Latin America. The Córdoba affair was also important within the frame of the history of science in Argentina, and has only recently been documented.[1] It contributed to the characterisation of scientists in Argentina as prone to becoming rebels, a perception that persisted over a considerable period of time in that country.

Eduardo L. Holmberg, Naturalist and Writer: the Two 1875 Novels

A keen supporter of the German scientists in Buenos Aires was a young medical student, Eduardo Ladislao Holmberg (1852–1937), then in his early twenties. He was already known as an upcoming naturalist, a leading specialist on Argentina's spiders and a fine writer. Together with palaeontologist Florentino Ameghino, from the 1870s he became one of the most influential scientists in Argentina. Within his generation, Holmberg was the most vocal fighter against prejudice and superstition, and a committed Darwinist.[2] By the end of the century, and mainly through the influence of Holmberg's literary inclinations, an old intertwining between the scientific and literary communities of Argentina was preserved and even deepened.[3] This was at a time when science had a special resonance in the world of literature.

Holmberg wrote two novels in 1875 in which local research scientists play, for the first time in the Argentine novel, a central place. The first of

---

1   Eduardo L. Ortiz (1994) *Los Seis de Córdoba* (Córdoba: Academia Nacional de Ciencias).

2   However, Holmberg could not be classified as an uncritical supporter of Spencer's social Darwinism, see Eduardo. L. Ortiz ([1982] 1984), Conferencia de Clausura del II Congreso Nacional de la Sociedad Española de Historia de la Ciencias: La polémica del Darwinismo y la inserción de la ciencia moderna en Argentina, *Actas, Congreso de la Sociedad Española de Historia de la Ciencia* (Madrid, vol. I), pp. 89–108; on p. 98.

3   See Eduardo L. Ortiz (1998) 'The Transmission of Science from Europe to Argentina, and its Impact on Literature: from Lugones to Borges,' in E. Fishburn, (ed.), *Borges and Europe revisited* (London: Institute of Latin American Studies), pp. 108–12, in pp. 113–4.

them, *Dos partidos en lucha*,[4] was a Darwinian fantasy placed in the Buenos Aires of the early 1870s. Burmeister, who opposed the theory of evolution, is a leading but distant character in the novel; however, nothing less than Darwin's opponent. Other characters in *Dos partidos* are either real, or have features related to some of the main young scientists in the Buenos Aires of the time. In 1957 Antonio Pagés Larraya published a most important *Estudio preliminar* to a collection of Holmberg's short stories.[5] Although this book does not include the 1875 novel, Pagés Larraya briefly referred to it and suggested that the main characters in *Dos partidos* are 'easily identifiable' with contemporary local scientists.[6]

The second of Holmberg's books is *Viaje maravilloso del Señor Nic Nac en el que se refieren las prodijiosas aventuras de este Señor y se dan a conocer las instituciones, costumbres y preocupaciones de un mundo desconocido. Fantasía Espiritista*.[7] It was originally serialised in the *El Nacional* newspaper, of Buenos Aires, from 26 November 1875, and printed in book form in 1876. While *Dos partidos* was called a scientific fantasy, *Nic Nac* is a spiritist fantasy. The spiritist Nic Nac is a counterpart of the materialist and Darwinist Ladislao Kaillitz of the first novel. The name of Kaillitz is made up from Holmberg's second given name and a variant of the original name of his family.[8] Nic Nac is the name of a brand of biscuits produced by the almost fully automated, high technology factory, built by the firm of Bagley & Co. in Buenos Aires.

Nic Nac is a much more complex, subtle and hermetic novel than *Dos partidos*. Difficult to read, and difficult to grasp, it has not attracted a great deal of attention in the limited literature on Holmberg. Pagés Larraya's notewor-

4   Eduardo L. Holmberg (1875), *Dos partidos en lucha, Fantasía científica* (Buenos Aires).

5   Antonio Pagés Larraya (1957) 'Estudio preliminar,' in E.L. Holmberg, *Cuentos fantásticos* (Buenos Aires), pp. 7–98.

6   In Pagés Larraya (1957), p. 53. Besides this last author pioneering work, *Dos partidos* has been discussed in Marcelo Monserrat, 'Holmberg y el darwinismo en Argentina,' *Criterio*, XLVII, 1702, 24 (1974) and in Ortiz (1982).

7   Eduardo L. Holmberg (1875) *Viaje maravilloso del Señor Nic Nac en el que se refieren las prodijiosas aventuras de este Señor y se dan á conocer las instituciones, costumbres y preocupaciones de un mundo desconocido. Fantasía Espiritista* (Buenos Aires). (Although serialised from 1875, it appeared in book form in 1876.) The story is also called *Viaje de Nic Nac al planeta Marte*, on p. 10. Today, this book is very rare, absent from the catalogue of the British Library, London; the Bibliothèque Nationale, Paris, the Library of Congress, Washington, or even the Commulative Catalogue of US Libraries. It is included in the edition in progress by Eduardo L. Ortiz (ed.) *The Works of Eduardo L. Holmberg* (London, in preparation).

8   His grandfather was Eduardo Kannitz, Baron Holmberg, in different documents his name is also written Kanitz, Kaunitz, Kaulitz and Kallitz.

thy five-page critical discussion,[9] which did not attempt to penetrate into the issues of the novel, has been the source of all subsequent remarks of later authors. Holmberg's allusion to spiritism, highlighted in the title of the novel, and transparent references to Argentine society, may have contributed to obscure the existence of a deeper, historical-scientific code in it.

A more subtle line in *Nic Nac*'s plot becomes visible when the novel is projected against the background of the *Córdoba Six* affair, that is, on the origins of the Argentine National Academy of Science. From this perspective, the mainly Córdoba-based *spiritist* fantasy becomes a meaningful, although much richer and complex parallel to *Dos partidos*, the Buenos Aires's centred *scientific* fantasy.

In 1878 Holmberg published a fragment of a third novel, of a similar character to the two 1875 ones: *El tipo más orginal* [*que he conocido*]. Although published in 1878, he also started writing it in 1875 and read parts of it at the literary society *Academia Argentina* in 1875 and 1876. The sections that have survived were published starting from the second number of the Buenos Aires literary magazine *Album del Hogar* edited by the poet Gervasio Méndez. The novel is not considered in this chapter.

II. Realism and Fantasy

Realism and Fantasy in the Argentine Literature of the 1870s

Although it has been stated by Pagés Larraya, and accepted by further critics, that 'there are no characters or types designed with a realistic criteria'[10] in *Nic Nac*, I wish to show that this journey, or dream, has been constructed on the basis of real characters acting out a genuine and recent dramatic event, the *Córdoba Six*, rather than being the product of pure fantasy.

This is not lacking in interest. Leading analysts of Argentina's narrative of the second half of the nineteenth century, such as Ricardo Rojas[11] and Pedro Henríquez Ureña,[12] have emphasised the naturalistic and realistic features of most of the narrative of the period, while noting some fantastic elements in Holmberg and some subsequent writers. More recently,

9   Pagés Larraya 'Estudio preliminar,' pp. 57–62.
10   'No hay caracteres ni tipos diseñados con criterio realista,' Pagés Larraya 'Estudio preliminar,' pp. 59–60.
11   Ricardo Rojas (1917–22) *Historia de la Literatura Argentina* (Buenos Aires), in vol. 4 (or vol. VII–VIII in the more recent edition *Historia de la Literatura Argentina, ensayo filosófico sobre la evolución de la cultura en el Plata*).
12   Pedro Henríquez Ureña (1945) *Literary Currents in Hispano-America* (Harvard, translated as: *Las corrientes literarias en la América Hispánica*, Mexico City, 1949).

Adolfo Bioy Casares,[13] Pagés Larraya,[14] Ana María Barrenechea[15] and Nicolás Cócaro[16] have returned to this topic. They focused more intensely on the presence of the fantastic, pointing to precedents in Argentine literature, and naturally making reference, explicitly or implicitly, to Holmberg's early contributions to the fantastic stream, without going into any detail on his novel *Nic Nac*.

Without entering into the complex problem of what exactly constitutes fantastic literature, I wish to draw attention to the fact that in *Nic Nac*, as in Holmberg's previous 1875 novel *Dos partidos*, scientific fantasy is still rooted in a realistic construction. That is, that these novels belong to a very definite group, in which fantasy is constructed over a concrete, recent, poignant and historically documentable situation, in which the realistic content is still substantial and drives the story.

As several authors have indicated,[17] a common feature of fantastic literature is extrapolation, that is, a rational projection of what science knows into uncharted territory. Karl Kroeber, in particular, suggested that science fiction appears when the supernatural is driven out of enlightened society.[18] Instead of looking for magicality, the consequences of the same scientific and technological advances that are driving superstition out are extrapolated. One can detect in this literature a desire for the anticipation of ideas and results that are not yet within the grasp of science, and that are channelled, instead, into literature or poetry.

This is not the case with Holmberg's two 1875 pieces. The main question in the two novels, strictly considered, is not a scientific one. The first uses a dramatisation of a local discussion on Darwinism, though not on the actual theory of Darwin, a topic on which Holmberg — an entomologist and not a biology theoretician — did not consider himself an expert.[19] The second uses spiritism as a vehicle to annihilate the mass sup-

---

13 Adolfo Bioy Casares (1940) 'Prólogo,' in Jorge Luis Borges, Silvina Ocampo and Adolfo Bioy Casares, *Antología de la literatura fantástica* (Buenos Aires).

14 Pagés Larraya (1957).

15 Ana María Barrenechea (1957) 'Introducción,' *La literatura fantástica* (Mexico City).

16 Nicolás Cócaro (1960) 'La corriente literaria fantástica en la Argentina,' in *Cuentos fantásticos argentinos*, selected by N. Cócaro (Buenos Aires).

17 Patrick Parrinder (1980) 'Science Fiction and the Scientific World View,' in P. Parrinder, *Science Fiction: its Criticism and Teaching* (London), pp. 67–88; Darko Suvin (1983) 'Victorian Science Fiction, 1871–1885: the Rise of the Alternative History Sub-genre,' in *Science Fiction Studies*, vol.10, no. 2 (1983), pp. 148–65; Kingsley Amis (1960) *New Maps of Hell, A Survey of Science Fiction* (New York).

18 Karl Kroeber (1988) *Romantic Fantasy and Science Fiction* (Yale).

19 Eduardo L. Holmberg (1882) *Carlos Roberto Darwin* (Buenos Aires), pp. 69–70.

port of the human soul, that is, as a convenient artefact to elude entering into the complexities of the mechanics of interplanetary travel. Holmberg discussed the impact of switching off some specific properties of matter in other stories; for instance, in *Horacio Kalibang*,[20] where the automata is able to play with the position of its centre of gravity.

Clearly, there were other useful sources to learn about Darwinism, spiritualism, the occult, interplanetary travel or the plurality of inhabited worlds in the Buenos Aires of the mid 1870s, even without going out of the Spanish language.

As Pagés Larraya already perceived, there are in *Nic Nac* definite statements on Argentine society. Social critique, however, can be found in several other Argentine works of the time. Perhaps, the application Holmberg makes of it to the discussion of the incipient and troubled Argentine scientific community of the mid-1870s, and of its relations with the wider world of culture in that country, has a deeper interest and a more clear originality.

No doubt science plays a role in his stories, often a subversive one, and gives them a more solid base, but at the same time the novels are used to deliver a message more closely related to the sociology of science than to science itself. In this respect, these most interesting scientific-spiritist-fantastic novels belong in the class of pedagogical devices with a scientific-sociological message. Because of this, they can be regarded as a literary formulation of the question of consent and resistance in relation to science within the much larger cultural community of Argentina at the time. In this perspective, the book has also an explicit propaganda function for the scientific community and tells us about the existence, in the 1870s, of tensions within it, as well as in its relations with the community of culture in Argentina.

Almost 40 years after the publication of *Nic Nac*, Holmberg gave an assessment of his own creations and stated that all his work, particularly 'those corresponding to the domain of fantasy' emerged from voluntary combinations directed by scientific, literary or artistic facts.[21] This statement could be no better applied than to his two 1875 novels and the unfinished one of 1878.

20   Eduardo L. Holmberg (1879) *Horacio Kalibang o los autómatas* (Buenos Aires). This work was dedicated to José María Ramos Mejía.

21   '... todas mis obras, cualesquiera que sea su carácter, y en particular aquellas que corresponden al dominio de la fantasía, surgen de combinaciones voluntarias dirigidas en su expansión o desarrollo por la razón y por los conocimientos científicos, literarios, y artísticos, sin los cuales no hay escritor de fundamento, y sin esclavizar por esto las espontaneidades de combinaciones que se realizan en el cerebro bajo el impulso del pensamiento en acción'. 'Un fantasma,' *La Cruz del Sur*, vol. I, no. 7 (Buenos Aires, 1913), pp. 331–46, on p. 332.

The Argentine cultural community of the 1870s was essentially a literary one. It is in literature that we find some of the highest levels of cultural achievement, continuing a tradition that began shortly before the middle of the century with Sarmiento's *Facundo*. It is then of some interest to look there for signs of the emergence of a scientific sensibility in the young Argentine intellectuality of the time. This chapter is part of a study in that direction; one of the most relevant sources, Holmberg's *Nic Nac*, is discussed here. I will make some remarks on Holmberg's particular selection of factual material; deal with the identification of the plot-line, relating it to little known events in the Argentine scientific community shortly before the novel was written; and, finally, make some remarks on its importance in relation to the discussion on the emergence of the scientific and the fantastic in late nineteenth-century Argentine literature.

III. Córdoba in the 1870s

Scientific Córdoba in First Half of the 1870s[22]

In September 1869 the Argentine Parliament approved a budget for the construction and support of a national astronomical observatory. Benjamin Gould was invited to become its director, to select personnel and to buy in Europe, and in the United States, the necessary books, instruments and ancillary equipment.

In September, Gould arrived in Argentina and, with a group of American collaborators, started work on mapping the southern sky. By the end of the year, and also in Córdoba, Professor Paul G. Lorentz, formerly at the University of Munich, was put in charge of the new chair of botany, while Dr Max Siewert took up the new chair of chemistry. Shortly after, a new member was added to the German scientific family in Córdoba: Carl Schultz-Sellack, an expert on astronomical photography, with a solid training in photochemistry in Berlin and an interest in American ethnology. With his help, the Moon was photographed from Argentina and, soon after, successful trials of this new technique on large scale stellar photography of the Southern sky began. Another professor, entomologist Dr Hendrik Weyenbergh, arrived in Córdoba with his wife and established a new chair on zoology; a new chair of mineralogy was

---

22   This section is based on Eduardo L. Ortiz (2001) *Sarmiento, Gould y la inserción de la ciencia astronómica en Argentina*, Academia Nacional de Ciencias, Inaugural Academic Lecture (Córdoba).

made available to students of the University of Córdoba by Dr Alfred Stelzner. Towards the end of the year a professor of mathematics, Christian August Vogler, a German expert in geodesy, was hired for the university and arrived in town. He was attracted by the possibility of measuring in Argentina an arc of meridian. This operation would contribute to resolve an important scientific question: the true shape of the earth, as he indicated in his inaugural lecture. In the meantime, substantial purchases were made in Europe, particularly in Germany, to equip the science laboratories and to considerably modernise the science contents of the old and rich university library.

Chemist Siewert wrote a textbook on analytical chemistry for the students of his course, which he had started in March 1871. At the same time, he was advising the government on a paper manufacturing plant. Siewert represents applied science in the group. In the meantime Gould had, ready for the printers, his *Uranometría Argentina*, a new catalogue of the southern stars, up to ten degrees north of the equator. Finally, it was expected that soon Burmeister, who had arranged the contracts with the foreign scientists, would move to Córdoba.

The need to modernise the university, and create new careers, particularly in the direction of the sciences and engineering, was often stressed in the daily press, and in Parliament, in Buenos Aires. The local press in Córdoba remarked that science studies had been the least important in the university's life in the past, but should come right to the fore now.

The scientific balance was tilting from Buenos Aires to Córdoba, where a good number of leading Argentine intellectuals of the time had been trained. Sarmiento's government had designated that city, considered at the time as a possible new capital city, as the cradle of modern science in Argentina. The government's intention was to establish an academy of sciences in Córdoba, with the most distinguished scientists available in the country including, of course, Gould and the German scientists in Córdoba as part of its membership.

Never in Argentina, before or after these few years, were so many scientific institutional advances made in such a short period of time. It was a journey into the unknown, or a dream that, sadly, encountered some difficulties. They reached literature.

The Córdoba Six

In December 1873 the government promulgated a decree creating a National Academy of Sciences in Córdoba on the basis of the project submitted by Burmeister. However, this project had been produced without consultation with Córdoba's German scientists, whose scientific activities were placed under Burmeister's tutelage and direct scrutiny. The scientists reacted sharply against the proposal, regarding it as an interference with their academic freedom, which had been guaranteed in their contracts. Besides, they objected to an academy directed by Burmeister from Buenos Aires. The example of the *Göttingen Seven*, who fought for academic freedom in the Germany of the 1830s, was now placed at the centre of the scene at the University of Córdoba of the 1870s.

Burmeister promoted a former minor chemistry assistant as his personal representative in Córdoba, and interim director of the new academy. Scientists perceived this as an affront and wrote a measured letter of complaint to the minister of education. This resulted in Lorentz's resignation, followed, in quick succession by those of Siewert, Stelzner, Weyenbergh and Vogler. A dimension of the loss could perhaps be measured by the fact that while Stelzner was in Córdoba, he was offered a chair at the world famous Mining School in Freiburg; he rejected the offer expecting to remain in Córdoba. The offer was renewed and he moved to Freiburg. Vogler also had a remarkable academic career in Germany, where he became a well-known specialist in geodesy.

The conflict reverberated until after Avellaneda had left the Ministry of Education to become a candidate, to follow Sarmiento as president. The new education minister, Dr. Juan C. Albarracín, rejected the terms of the letter sent by the German scientists, and also their resignations, and expelled them all.

Burmeister's standing as a scientist, and his substantial contributions to the knowledge of the natural science of the Argentine territory were never under discussion. However, many deplored his conflictive personality. The blind support of the Argentine authorities for Burmeister caused a deep impression in the country and abroad. After Burmeister was finally forced to relinquish his hold on the academy, it somehow recovered, but not without paying a very heavy price in terms of scientific manpower. As Holmberg pointed out nearly half a century later,[23] the *Academy* proceedings were one of the most positive remnants: its pages became the vehicle for the communication of the research work done in Argentina by a generation of young naturalists.

---

23 In 1922, in Eduardo L. Holmberg (1922) *Historia de la fauna argentina* (unpublished manuscipt, Archivo de la Academia Nacional de Ciencias Exactas, Físicas y Naturales, Buenos Aires).

Argentina's Instability at the Time of Holmberg's Two Novels

The early 1870s were extremely hard in Argentina. After the end of a bloody international war, former president Mitre's uprising against the newly-elected successor to Sarmiento led to an internal military confrontation in 1874. In 1875, the year in which both of Holmberg's novels were written, the Colegio del Salvador, the elite Jesuit school in Buenos Aires, was burned to the ground. Although this outrage was generally attributed to the work of émigré Italians, followers of Garibaldi, more local participation was not difficult to trace. In addition, Buenos Aires had been visited by yellow fever in the summer of 1871, an epidemic that took a heavy toll in the capital city and almost paralysed the national administration.

This was also a period of a large expansion in exports, as well as of scientific and military exploration of the physical space of Argentina, when a modern conception of nation was being developed. Only a few years later, in 1879, the government of Argentina attempted a local version of the American *Conquest of the West*, the so-called *Conquista del Desierto*, which reclaimed for possible European immigration land traditionally occupied by an indigenous population. Within a few years four important works dealt with this question from different angles. Two of them, from a literary perspective: Lucio V. Mansilla's *Una excursión a los indios ranqueles* and José Hernández's *Martín Fierro*. The third, by Alvaro Barros, *Fronteras y territorios federales de las Pampas de Sur*, was a military analysis of the situation of the frontier with the Indians. The fourth was a natural history of the territories visited by the army, prepared by a group of German naturalists that accompanied the expedition, some from the reconstructed academy; young Holmberg contributed to edit their observations.

At the same time, by means of the telegraph, steamships and railways, the virtual physical space of Argentina was contracting fast, helping it to become a more functional nation, with a faster central line of command. Time distances with the world outside were also becoming much shorter. The submarine cable removed time-lags: Argentina ceased to live in the history of history, taking the pulse of the present, as events happened. Its commercial affairs became more closely linked to the needs, demands, and also permanent oscillations, of European markets.

Let us now go back to *Nic Nac*, and consider the story in a different reading.

III. The Novel

From Plaza de la Victoria to other Inhabited Worlds

In the 1870s there was in Argentina some fascination with exploration. The novel *Nic Nac* reflects it with a long journey into uncharted lands. As with the earlier novel of Holmberg, *Nic Nac* starts with a real, or supposedly real, situation, from which the action gradually gathers momentum into the realm of fantasy. The novel opens at the Plaza de la Victoria, in Buenos Aires, on an evening towards the end of November 1875, where groups of people discuss the return of an explorer named Nic Nac. The author gives some interesting details on Buenos Aires's daily life, makes reference to Mitre's recent uprising and to the burning of the Jesuit school. He makes ironic remarks on Gould and his observatory in Córdoba, and also on the medical profession and the attempts of some of its members to understand human psychology with the help of physiognomy, phrenology and even with a look into the occult.[24] Real characters, like Dr. José María de Uriarte, director of a mental institution, and Holmberg's eminent friend and mentor, Dr. José María Ramos Mejía, are mixed with others in which imagination has a definite hand.

These brief introductory remarks end with news that a few days later a book with memories of Nic Nac's journey had been published in Buenos Aires.[25] Holmberg reproduces it in the novel. The memories are written in the first person, but quoting extensive dialogues. With the news of Nic Nac's book, the dailies also reported that the author had been admitted to Dr Uriarte's hospice, where he was given a treatment with cold showers, possibly to extinguish the fire of his imagination.

---

24 A number of books were then available on these subjects. On occultism Eusèbe Salverte (1865) *Las ciencias ocultas. Ensayo sobre la magia, los prodigios y los milagros*, introduction by M.E. Litré (Barcelona), 446 pp., was one of the best. The Spanish translation was made from the French third edition of Salverte's *Des Sciences occultes, ou Essai sur la magie des prodiges et les miracles*, a book developed following the author's earlier studies on the history of civilisation. Translations had already been published in London, in 1846, and in New York, in the following year. In other, later, fictional work Holmberg made reference to Louis Jacollicot; his nearly 400-page work on the origins of Christianity, and on Brahhamanism: *La Bible dans l'Indie. Vie de Iezus Christna* was published in Paris, four editions (from the fifth to the eighth) were printed between 1875 and 1876. It was also translated into English in 1870, in London, and two years earlier into Spanish by R. Comás Solá: *La Biblia en la India. Vida de Iezeus Christna*, in two volumes, for the then well-known *Biblioteca Contemporánea*, Barcelona.

25 The title of Nic Nac's memories is more direct: *Viaje maravilloso del Señor Nic Nac al planeta Marte*.

Nic Nac is an inquisitive man of nearly 40, almost twice the age of the author of the novel. At the age of 20, wondering about the reality of his black cat, he began considering the interplay between reality and fantasy. Lured by the teachings of Friedrich Seele (or Federico Alma), a visionary, a sage and a medium recently arrived from Germany, Nic Nac travelled to a distant planet and returned to Earth to tell the story of his extraordinary journey. His journey was part of a philosophical experiment to learn more about terrestrials; however, his experiences there mirrored some of those he had already tasted on Earth.

This story within a story opens with a critique of academic philosophers, whom Nic Nac accuses of being ignorant of the recent results of scientific research, and follows with a lively defence of the experimental method. Interestingly, there is also ironic treatment of positivism, the new philosophical doctrine that was to stay in Argentina for a very long time. Its acceptance by leading Argentine scientists has, perhaps, been exaggerated. Nic Nac tells us that when local 'positivists' visited Seele they perceived his talent but, unlike the independently minded Nic Nac, ideological prejudice forced them to reject the potent ideas of the visionary.

Nic Nac's Planetary Passions

From the start, we are relieved from any concerns as to the mechanical feasibility of the interplanetary journeys. Unlike Verne's sophisticated mechanical arguments, in Holmberg's story it is not the weighty body that makes the journey, only the imponderable soul. Seele maintained it was possible to break the link between matter and spirit by experiencing with some kind of fakirism: fasting to death. Nic Nac agrees to start a journey of planetary exploration that will require his physical death, but also the liberation of his spirit. Conquering truth through suffering is emphasised in this novel as well as elsewhere in Holmberg's writing.

Some circumstances surrounding Nic Nac's physical death impress so deeply the doctor certifying his death that he too dies. This sensitive individual is emblematic of romantic love in the novel. The souls of the two men mix up and disappear into space, only to find, in their timeless wanderings, Seele's spirit. This is the only point in the novel where the author plays with time in dissociation with space.

They finally arrive on a planet, Mars, where they discover, to their surprise, that it has physical characteristics entirely similar to those of Earth. Later, we learn that the parallels go beyond geography. They have died and travelled only to return to a reality that is not dissimilar to the one they left behind on Earth: Nic Nac and the doctor have not been able to escape the peculiarities of humanity.

After some experiences, a ceremony of Marsification took place at a cave in the Nevado de Famatina. The ceremony was presided by a genie of the mountain, who gave the two friends a phosphorescent glow. They were also reunited with their bodies, and a part of their terrestrial memory. Later, Nic Nac discovered that the mountain genie was none other than Seele, who explained to him that he, as well as being Nic Nac's feline companion on Earth, actually belonged to Mars.[26] From now on Germanic references to Seele are dropped in the novel.

The character named Seele follows an irregular pattern in *Nic Nac*; possibly, the last sections of this novel were written when the first ones had already appeared in *El Nacional*. Seele transmigrates in the novel from a visionary on Earth to genie of the mountain on Mars, but in different parts of the books is also referred to as a sage, the black cat's companion and the embodiment of perfidy. Seele approaches and comes away from definite real models. In large sections of the book Nic Nac recognises him as his master, as much as Holmberg recognised Sarmiento as his mentor. The complexity and ambiguity of this character also suggests that an idealised Burmeister, deprived of all his personal miseries, may also have contributed to build some facets of it.

Throughout the story the black cat is associated with inquisitiveness; in Mars it becomes their animal magnetic compass.[27] Helped by this intelligent animal that was the philosophical cause of Nic Nac's travel, and of the disruption of his life, the two terrestrial visitors began to explore the new planet. They are in a territory that later revealed itself as a Martian parallel of Argentina, there called Aurelia, replacing silver with gold. In later work, Holmberg used the figure of translating Argentina's reality to a different country or world, to make amusing, and often acid, remarks on his countrymen and their society.[28]

26 Holmberg (1876), p. 38.

27 *Ibid*, p. 46.

28 For example, in Holmberg's *Olimpo Pitango de Monalia*, a manuscript finished around 1915 and recently edited with an introduction and notes by Professor Gioconda Marún (Eduardo L. Holmberg, *Olimpo Pitango de Monalia*, Buenos Aires, 1994). The autographed manuscript of this work is reproduced in *The Works of Eduard Eduardo L. Holmberg*.

Theosophopolis

The black cat invites them to travel in a south-eastern direction; they finally reach a large city crowned by high towers. Its name is Theosophopolis, the city of God and the wise men. It has two remarkably different parts. The east side, called Theopolis, is sad and lugubrious, with a depressing sound of bells; in addition, Theopolites wear dark long dresses and share in the sombreness of the place. Sophopolis, on the contrary, is animated and bright; its people had a rosy complexion, were talkative, and full of life.[29] There was little contact between the two sides.

As night fell, Nic Nac and the doctor noticed that artificial light was unknown in Theosophopolis. As its firefly inhabitants[30] have their own phosphorescence they did not feel the need to invent artificial light, which Nic Nac praises as a 'civilising element', and one of the great conquests of humanity.[31] We later learn that the colour and intensity of the halo changes with external as well and internal circumstances.

Once in the city, the black cat takes them into a long procession, where they first meet inhabitants of Theosophopolis. A chorus sings in an unintelligible monotone language, while an amiable Sopholite approaches them and becomes their cicerone for the rest of their journey. He explains to them that when Sopholites become old they move to Theopolis to take transmigration, a form of evaporation performed on them by the intervention of the Great Priest. This ceremony takes place at a special space, of prismatic form, in Theopolis's *Temple of Regeneration*.

As the body of the Sopholite is volatilised, it fills the prism at the centre of the *Temple*; his soul flows into the ether, to an undetermined planet. When this happened, Nic Nac and the Doctor noticed someone close to the prism, with a notebook and a large matrass, a large rounded glass bottle with a long neck and a graduated tap. They overheard the man with the matrass, whom I shall call the Chemist, saying that knowing the amount of oxygen in the gas produced by the evaporation of a Sopholite, it would be possible to estimate the amount necessary to burn down the neighbourhood where the Theopolites lived, together with all its inhabitants.

---

29   Most of the scientists, and also foreign emigrants, lived on a side of the city away from the main historic buildings of Córdoba.
30   Holmberg (1876), p. 80.
31   *Ibid.*, p. 51.

The Evolution and Mars

Nic Nac says little about Martian eschatology. The point of destination of transmigrated souls is elusive: nobody knows for certain to which planet a soul would migrate. Reflecting on uncertainties, Nic Nac remarks that on Mars, as on Earth, not much is known. The more fundamental questions: what is life?; what is the ether?; what is electricity?; who are we?; where do we come from?; where do we go? all remain unanswered. 'I don't know,' he says, 'seems to be a colossal giant separating humanity from truth.'

Evolution could not be absent in a book by Holmberg. Arguing from the angle of the theory of the origin of the solar system, that is, on evolution applied to the life of planets,[32] Nic Nac estimates that Mars is an older planet than Earth, and that life on it must have emerged earlier. However, its evolution has been slower.

Nic Nac returned to a discussion on social incentives for civilisation and for the need to invent, and in this connection introduced the question of the emergence of religion. The Martian's phenomenon of personal glows is also explained in terms of natural selection: since Mars has no Moons, nights are pitched dark there, and consequently hard for those without a phosphorescent glow.[33] But Nic Nac argues, dialectically, that the aura has also been the cause of the slow evolution of Martian society, it has deprived them of the incentive to invent artificial light. Nic Nac concludes that civilisation is the daughter of a struggle established between human beings and the natural elements: The stronger the struggle, the more powerful the resulting civilisation. Is phosphorescence telling us something on the effect of wealth on the society he left back on Earth?

He also concludes that the education of intelligence must be in harmony with the sum of human necessities. Later, his friend the cicerone would challenge this mechanical view of intellectual evolution.

Two Kinds of Christian

The cicerone explained that initially, the city was only inhabited by Sopholites, who were Christians. Later, others requested permission to settle on the east side of the city. The new inhabitants were also Christians, but 'transformed Christians'. The moral physiognomy of these men is not encouraging. Besides, they are described as being members of a family rejected in every country, who had made hypocrisy their dogma. The

---

32 *Ibid*, pp. 61–2.
33 *Ibid*, p. 83.

Theopolites began with temporary tents, but soon built their 'severe edifices' in town.[34] Thirty years later the city was renamed Theosophopolis. There are some suggestive coincidences in the chronology: Jesuits were allowed to return to Argentina during the government of Juan Manuel de Rosas, in the mid-1830s.

Trouble between the two groups started with the abduction of some beautiful Sopholite women by the Theopolites. Asking for reasons for the abduction, the cicerone explained to Nic Nac that Theopolite women were unattractive, squalid, yellowish, some lacking an arm or a leg. They were of no better quality than an image of Jesus Christ they saw in the *Temple*. After the abduction, a more sophisticated generation of Theopolites emerged; in them, hypocrisy mingled with refinement.[35] The appropriation of women's minds through the manipulation of their education was a topic of serious concern in the Argentina of the mid-1870s. Not only liberals, but also other religious orders, accused the Jesuits of such action.

The Sophopolis Academy of Sciences

The cicerone invited them to a meeting of the Sophopolis Academy of Sciences; on their way, they crossed with a group of men arguing and shouting at each other; to Nic Nac they behaved like a pack of wolves. The cicerone explained that they were a zoologist, who specialises in amphibious creatures, and an astronomer, interested in asteroids. Their homes faced each other, and the zoologist complained that the tube of the telescope disturbed his animals, while the astronomer complained that water condensation from the animal's pond affected the lenses of his instrument. Nic Nac's reflection on personal relations in Sophopolis's community of science was pointed: 'poor savants, everywhere the same, always bad tempered and often impertinent'.[36]

Although the zoologist Weyenbergh did not live near Gould's home in Córdoba, up at the Observatory's heights, his rented home was also his zoology cabinet, as the university did not yet have rooms specially arranged for that purpose; he and his wife resided in the company of his beasts. Gould and the so-called German scientists did not openly quarrel, but he deeply mistrusted them, as his personal correspondence reveals. Scientific academies were not new to Gould, who was one of the founding members of the United States National Academy of Sciences; however, when the Argentine

34   *Ibid*, pp. 73–4.
35   *Ibid*, p. 78.
36   *Ibid.*, p. 82.

Academy of Sciences was established, mainly made up of German scientists residing in Córdoba, he refused to honour it with his membership.

A section of Swift's travels[37] is a parody of scientific research at the Royal Society, in London, founded half a century before. In a similar Swftian spirit, a group of scientists investigated freely at the Sophopolis Academy of Science, trying to reveal the secret laws of morals and physics. The Sophopolis corporation was divided into two sections: pure and applied science but, as in Córdoba, the former was perceived as the source of all progress.

The discovery of harmonies in nature was regarded by the academy as one of the characteristics defining humanity. This view is emphasised in the novel and contrasted with Nic Nac's terrestrial, more pragmatic and utilitarian views on scientific research. The cicerone also accepted evolution as well as the possibility of an 'evolutionary' view of culture, but he also thought man's intelligence was too precious to be confined to the satisfaction of the basic necessities of ordinary life.

A Stormy Session at the Academy of Sciences

The large Academy's hall was illuminated by the aura of the sages; Nic Nac remarked: 'There, science irradiated its light.'[38] He was introduced to astronomer Hacksf, in whom traits of Benjamin Gould's personality can be detected. The Zoologist Biopos, and the Botanist Geot are fictional characters in which elements of at least two contemporaneous German scientists in Córdoba: Dr Weyenbergh and Dr Lorentz respectively, can be detected, particularly of the first of them.

The Chemist, the strange and impulsive man of the matrass,[39] was also there, seating with the applied scientists. Nic Nac sat with the theoreticians, next to Hacksf, who asked him about the types of telescopes used on Earth. He also enquired on the abundance of zoologists there, possibly contemplating a move to Earth if zoologists were less conspicuous over there.

The topic under discussion at that session was a review of a *Complete Treatise on Marsography*, a physical description of Mars. This was also a controversial topic at the time in Argentina. The need for a physical description of the country was clear; besides its scientific interest, it was essential

37   Jonathan Swift (1726) *Gulliver's Travels* (Dublin), part II, chapter V.

38   'Allí la ciencia irradia luz,' Holmberg (1876), p. 99.

39   The biographies while in Córdoba, in the early 1870s, of the two leading chemists of the group, Siewert and Stelzner, as well as that of assistant Doering, are still too imperfectly known to support speculation on analogies with the Chemist.

to attempt to integrate Argentina's natural resources into the world's markets. However, Holmberg and Burmeister had clashed on practicalities. While Burmeister wished to do it all by himself, Holmberg argued that for such an extensive piece of work a group of well-trained Argentine naturalists, some already known in Europe, should also be engaged. A discussion on the relative merit of scientific studies made in Tedecia, Ingelia, Gandalia, Spondia[40] and Tarantelia (clearly Germany, England, France, Spain and Italy, respectively) confirms, once more, that the tensions between different national groups existing in society in general in the Argentina of the mid-1870s had also permeated its scientific community.

At some point in their deliberations the academicians spotted a Theopolite who had sneaked into the academy's hall. This was a rare event, as they were not regulars at the academy's meetings. Using the fact that Theopolites were said to be *elastic*,[41] the Chemist pushed him inside his large matrass. At the same time, he announced that now he would be able to measure the oxygen required for the evaporation of a Theopolite, and then to estimate the amount required to burn Theopolis and all its inhabitants and free the country of the elements rejected in all countries, sending them out to the confines of space. A discussion on Christian charity, between Nic Nac and the cicerone, followed.

The cicerone remarked that the suggestion of the Chemist was unnecessary, as Theopolites were *inflammable*[42] characters. A new quarrel erupted among the scientists on account of the genie in the bottle. Biopos wished to conserve the Theopolite in alcohol, and add him to his zoological collection, but the astronomer, who wished to verify the flammability of the Theopolite, warned he was going to burn the Theopolite himself using a ray of Sunlight he had managed to condensate in a crystal, which he then produced. The botanist Geot, and the sensible president of the academy intervened; the latter stopped Haksf warning it was dangerous to play these tricks inside the *Academy* building. He also warned that when the time arrived, everyone could have a Theopolite, if he wished.

A New Journey: to the Capital City

At the cicerone's home Nic Nac and the Doctor met his three daughters and the Doctor fell for one of them. After some interesting alternatives, he married the beautiful Martian girl.

40   Later called Espondia.
41   In the sense that they show themselves as flexible, or adaptable, in their views. It is a direct reference to Jesuit education.
42   Again, Holmberg is playing with words.

If the ray of sunlight was a dangerous weapon, we are told that, properly used, it also was a ray of life for the Theopolite in the matrass. For the wedding, the astronomer presented the couple with a box containing the crystal with the versatile ray; which also had the property of uniting people in love. As soon as the selfless Doctor knew it could save the life of the prisoner, he refused to keep it and tried to bring the crystal close to the matrass. Suddenly, Seele intervened, stopping him and evaporating the Theopolite. Nic Nac could not understand what Seele was doing but, later, he was told that saving the Theopolite would have been dangerous for Sophopolis's future. Meanwhile, without anyone noticing it, the Chemist managed to collect some of the gas produced in the evaporation: he had obtained a sample and could measure the oxygen relation.

While the Doctor was on his honeymoon, Nic Nac and his master Seele travelled, across a vast plain, to the capital city of Aurelia. They left their bodies and only the souls travelled. As Pagés Larraya has already remarked,[43] some of the finest literary pages in Holmberg's book are to be found in these passages. In a chapter called 'Insomnio', Nic Nac makes a beautiful literary evolutionist progression from the mineral water of a torrent, through plants and animals, to man. He also expressed inexplicable feelings of longing and melancholy while travelling over terrain reminiscent of the one separating Córdoba from Buenos Aires. His partially recovered memory did not allow him to understand his feelings.

On route, Seele and Nic Nac discussed the character of the population in the capital city. The question of cosmopolitism, and its impact on the sense of nationality of its people, was one of the topics under discussion. The role of foreigners was also discussed; Seele objected to the tendency of immigrants to remain in the large cities, rather than moving up country and helping to extract its buried treasures, or work its fertile lands, which had been the purpose of attracting them. These considerations made him reflect on the failure of the Aurelian agrarian immigration policy. He also objected to foreigner's interference in the internal affairs of the capital city.

The master made it clear to Nic Nac that he was not visiting the capital to change it, but to study it. Furthermore, he advised him to keep away from politics; to study intensely; to make suggestions, but always let people do what they felt was best for them. To make the visit even more objective, reducing the chances of the observer affecting what was observed, Seele offered Nic Nac the chance of becoming invisible, as the author of the novel, which he accepted. The understanding that our knowledge of the external world is

43   Pagés Larraya (1957), p. 61.

conditioned by the presence of the observer,was an advanced epistemological viewpoint, which shows again that philosophical positivism was not accepted uncritically by scientists in the Argentina of the mid-1870s.

Their city is described as dominated by positivism; its people enjoyed a
golden yellow glow instead of Sophopolis' rose colour. It would be a mis-
take to interpret 'positivism' here in a philosophical sense; this word was
often used in that period, to mean the pragmatism of a philistine society.
Residents of the capital city are seen as agitated and turbulent but, at the
same time, generous and capable of offering Nic Nac a few surprises.[44]
Seele noticed in them a certain tendency to rapidly divide into decidedly
opposite camps: to generate *Dos partidos en lucha*.

However, the tendency to group into hostile camps seems to have been
more general in Aurelia than Nic Nac wishes to tell us. The reader has already been informed of even more hostile confrontations outside the capital city: Theosophopolis is a divided city; in its academy Theopolites are
strangers. The fact that most scientists in Sophopolis were foreigners is
overlooked, had this been taken into account, the question of interference
by foreigners may have required reconsideration.

The topography of the two cities differs vastly; instead of two segregated
areas, as in Theosophopolis, he found far more integration in the capital city:
A rather curious juxtaposition of mansions and precarious dwellings, large
temples and narrow irregular streets. Nic Nac also found substantial differences in their demography. The indigenous element, Nic Nac stated conclu-
sively, is in a minority in the capital city; its incidence in Theosophopolis is not
made precise, but we must assume was more substantial.

By the time this novel was written, Holmberg had made a few natural
history trips inside Argentina. In one of them, to the old city of Carmen
de Patagones, he followed Darwin's steps, fictionally claiming in *Dos partidos* that he had found relics left by the naturalist in the cliffs carved by
the Negro River that abruptly separated the Pampas from Patagonia. In
this journey, for the first time, Holmberg came into direct contact with the
Indians of the Argentine south for whom he kept, for the rest of his life,
a deep respect that was fully reciprocated by them.[45]

44   Holmberg (1876), p. 141.
45   In 1910, to celebrate Argentina's first century of independence, Holmberg wrote a
      long poem, *Lin-Calél,* in which he celebrated the Argentine Indian's epic. When
      Holmberg died in 1937 representatives of Argentine Indian communities paid their
      respects.

Fireworks

In a letter from Theosophopolis, the Doctor told Nic Nac that he was in serious trouble. Nic Nac persuaded Seele to return immediately, and asked for his help. On arrival they realised a great catastrophe was taking place there. This is a grand, tragic finale for a fantastic dream.

The Theopolites abducted the Doctor's wife, and successfully repelled the Sopholites when they attempted to rescue the young woman; Nic Nac found Sophopolis in flames. It was an immense *evaporation* that happened without the Great Priest's intervention.[46] Badly burned, the black cat was waiting for his friends, the reality of life in Theosophopolis had hit him hard. In the square, near the flames, Nic Nac found his friend the zoologist Biopos, and also Hacksf. The astronomer kept his composure.

Gould did not leave Córdoba on account of the conflicts; although he also was a state functionary, his position was less dependent on Argentine government affairs than that of the German scientists.

The Chief Priest, who Nic Nac reminds us again was 'the only one worthy of veneration'[47] among the hypocrites, contemplated the scene with disbelief. Misinterpreting a gesture, or an attitude, as 'an act hostile to his dignity', he evaporated the Doctor and several others with him. Hacksf warned Nic Nac that the Chemist was about to commit a supreme act of vengeance, evaporating himself to burn the side of his enemies: his evaporation produced a circle of flammable vapours around Theopolis to avenge Sophopolis. All died in the flames of the great fire. As the story ends, Seele, the spirit of enquiry, reappears to condemn Theopolites to an eternal night. With Mars and Earth in opposition, Nic Nac's soul reached Earth; on the journey back he completely recovered his partially blocked terrestrial memory.

Irrespective of Burmeister's machinations, in the Córdoba affair Sarmiento and his minister Dr Albarracín were, ultimately, the signatories and the individuals responsible for the order to expel the German professors from Córdoba. The scientists had written a polite letter to the minister of education making remarks on Burmeister's plan, drawn up without consulting them. They had the right to do so as it changed radically the status agreed by the professors with the Argentine government in their contracts. The new rules compromised their freedom to undertake research, which sparked references to the *Göttinger Sieben*. Although the letter was a gracious petition, and not an act hostile to the dignity of the government,

---

46   Holmberg (1876), p. 177.
47   *Ibid.*, p. 181.

the authorities misinterpreted it as an affront. Evaporation followed; at the time it seemed to some that these actions had condemned Córdoba to an 'eternal night'. With this episode Nic Nac's book of memories ended.

Back to the Editor

At this point in Holmberg's novel the author takes control again. He tells us that Nic Nac's travels to another world were part of an internal journey, a state of ecstasy in which he wished to unravel the mysteries of humanity. It was expected that his extraordinary journey would throw light on some relevant questions. He remarks that, sadly, this had not been the case. The plurality of the inhabited worlds,[48] argued by Flammarion in Europe, had been proved, but it had only served to show that other worlds are not different from ours: our troubles seem to be inherent to human society.

However, it would not be fair to ignore the fact that, on account of his journey, Nic Nac had an opportunity to point to a number of interesting questions. He poked fun at popular customs and on the character of Aurelia's population in its two main cities, discussed social and emigration policies there; teased the medical profession; offered an insider's view of tensions within the local scientific community; debated on the relative merits of pure and applied scientific research; on evolution theory and its limitations outside the field of natural sciences, in particular in relation to spiritual life; on religion and education and; finally, on the interplay between invention, social needs and progress. He described Mars's natural scene with beauty and revealed the marked opposition that existed between Theosophopolis and Aurelia's capital city. Mars is also a natural scenario to throw irony on recent attempts to make of the Argentine army a professional career. In *Nic Nac* Holmberg also makes explicit analogies between the path of scientific research and a journey into the unknown with the help of imperfect instruments, on a road with multiple bifurcations where decisions need to be taken constantly.

He also tells us that tension in the intellectual world of Theosophopolis was such that a catastrophe was waiting to happen. The academy was placed in Theosophopolis because this was a city with an old and still solid cultural tradition but, at the same time, it was placed in a land of scientific infidels. No doubt the idea of Aurelians participating from Theosophopolis in the international quest for a rational understanding of nature, through the lenses of large telescopes and by doing experiments in

---

48   *Ibid.* p. 157.

modern laboratories, was commendable. Furthermore, sooner or later, advances would have also worked for the welfare of people, at least, through the much modern training that was to become available. However, Nic Nac is suggesting to us that it was misplaced; it had been located too close to the fire, and it soon became involved in a dramatic affair that was not of its own making.

Scientific Córdoba in the Early 1870s

No doubt widening the outlook of the old university of Córdoba unlocked reservations and uncertainties. There was unease, among some, as to the nature of the teachings to be inaugurated by the German professors who, in addition to teaching new subjects, were not necessarily Catholics. Their teachings offered an alternative to law and theology, which had been the university's traditional subjects since the 1620s, when it opened its doors. But, even if sectors of the traditional local elite of Córdoba may have been divided on the matter of enlarging the scope of the university, and in fact some expressed reservations of various kinds, which are even recorded in the proceedings of the university council meetings, it would be wrong to say that they were wholly hostile to it, or motivated exclusively by ideological reasons.[49] It must be borne in mind that these reforms were carried out, most forcefully, by graduates of the University of Córdoba, such as Avellaneda, placed then in high government positions. In addition, they were strongly supported by their progressive friends in Córdoba, who had an audience large enough to support publication of a leading newspaper there.

It was clear to many in Córdoba that careers in science would offer some definite advantages to a group of the students attending the university and, in the long run, may help keeping Córdoba's dominant position in Argentine university life. The fact that the university's rector gave these reforms his support, unreservedly, and that some young members of Córdoba's social elite joined science courses with enthusiasm, suggests support inside the university was not small.

Attributing an exclusive cultural meaning to the demise of the academy, or formulating it exclusively in religious terms, is not an option in which the Nic Nac story and contemporaneous historical evidence seem

49 Differences in salary between local and foreign professors was a more serious point of concern than ideology, as suggested in debates registred in Actas de Sesiones Consejo Superior, 1871–76, Acta 10, 1872; Archivo de la Universidad Nacional de Córdoba, Córdoba.

to converge. No doubt the dominant role of religion in Córdoba, and of the Jesuits in its intellectual life, was an element of resistance in relation to areas of modern science in the Córdoba of the time. Resistance operated directly or, in more subtle ways, through the high profile of distinguished university personalities. However, it should not be overestimated, for perhaps it was not necessarily the most important element, even if it may have been the most visible one, or the one most often invoked.

Besides, there is evidence of serious interest in modern science on the part of individuals in the University of Córdoba well before the arrival of Gould or the German professors there. If fact, a student from Córdoba gained a degree in mathematics, physics and astronomy at Göttingen University, nearly 20 years before the arrival of these scholars and helped them generously through his high political influence.[50] It was the university's fault if it did not take better advantage of his training.

The causes of the Córdoba affair were far more complex than Nic Nac assumed for Theosophopolis. The catastrophe at this city and, in particular, at the Sophopolis academy, speaks of sectarianism, of mutual hostility and of intense infighting, all of which existed also in Córdoba. But there the key decisions were not directly in the hands of the main actors, as the novel suggests for Theosophopolis, but in those of the national authorities, in Buenos Aires. It is interesting to remark that the author felt the need to attempt to justify in the novel the sharp reaction of the Chief Priest, without ignoring his share of responsibility in the tragedy.

The ambiguous role played by Burmeister, accepted as the government's advisor on sciences by the national political elites of Argentina, is also of some interest. It seems that, at a given point, he felt his authority to be eroded by the expansion of scientific foundations encouraged by Sarmiento's government outside Buenos Aires, and used his own political influence to apply the break.

There is no doubt, however, that Burmeister did not have sufficient executive power to act alone. The seriousness of this case, and the immediate history of the institution, once Burmeister was removed from the scene,[51] may well be telling us something about the tension between Sarmiento's good will and hard-pushed higher education reform policy, and the contradictions and ambiguities that existed inside the cultural, the scientific and the political communities of Argentina, and also between them. In fact, one minister of education, Albarracín, undid the work of the previous one, Avellaneda, who had already been elected president of Argentina.

50   See Ortiz (2001).
51   This point is considered in Ortiz (2001).

Besides serving as a dress rehearsal for showing Argentine cattle, agricultural and mineral products to the world, Córdoba's 1871 *Exposición Nacional*[52] was intended to show the relevance of new technologies for the development of agriculture, and also for some ancillary industry in Argentina. However, in the novel Nic Nac does not pay any attention to technological advances in Mars. Holmberg only makes some theoretical arguments on the rise of technology when discussing the phosphorescence of Martians, but cleverly eludes technology in connection with his journey.

There is little doubt that a polarity between science and technology existed in Córdoba at the time, and indeed in Argentina. A debate on policies on science and technology reached the pages of some of leading literary journals when two early scientific journals, each one reflecting one of these two trends, started publication in Argentina. This was immediately before the publication of Holmberg's first novel, which makes explicit reference to these periodicals. Although the academy was defended by wide sectors of opinion in Córdoba it is possible that a faculty of engineering, rather than a scientific institution, would have pleased much larger sections of the population.

If Holmberg's position in this debate favoured pure scientific research, no specific scientific theory or discovery is central to Nic Nac's story, as Darwinism was the reference in his earlier novel *Dos partidos*. Science is a general reference, often a metaphor to liven the plot. At the same time that science adds respectability to Holmberg's story, it gives a peculiar humanity to some of the odd characters of Sophopolis scientific community. It helps to convey a feeling of the complexity of the life of a scientific community, even of one that is only emerging. There is another, powerful, message in the novel, and that is that science is not an independent entity in Mars. As on Earth, its life and development is deeply intertwined with the development and pains of its contemporaneous society.

In addition to the enlightenment the novel provides regarding the sociology of science in the Argentina of the mid-1870s, its very existence seems to be justified in terms of the academy's enemies. They are as important, and prominent in the novel, as is science in Sophopolis. In this respect, the novel carries a message not only of juxtaposition and difference, but also of the sharp boundaries between communities engaged in separate monologues. This was also true of political factions in the Argentina of the time.

In addition to reporting fictionally on a piece of a real and unhappy event, central to the history of the insertion of modern science in

---

52   The full title was *Exposición Nacional de Artes y Productos Argentinos de Córdoba.*

Argentina, extraordinary travel and love are also components of this novel. Interplanetary travel adds to the fantastic component. Buenos Aires readers knew it from original sources, mainly from Verne. Spiritism, also dear to Buenos Aires audiences of the time, is reinforced by quotations, on topic and word, from astronomer Camille Flammarion who, it must be emphasised, was a well-known and committed spiritist. Love and courtship, less prominent than in *Dos partidos* and represented here by the Doctor, gives the novel a more human component.

Besides clear thematic analogies, in this as in other work by Holmberg, the influence of Edgar Allan Poe and above all of Ernst Theodor Amadeus [Wilhelm] Hoffmann permeates through the mixture of real and naturalistic scenes juxtaposed, without loss of continuity, to supernatural or fantastic events. Characters and situations in the novel are reminiscent, if often considerably differentiated, from Hoffmann's fiction.

A number of incidents, and characters, in this novel of Holmberg admit several levels of interpretation. As in some of Hoffmann's work, the action takes place in two parallel worlds, in which the writer's friends occupy what is possibly the higher level. The Keplerian blurred differentiation between animal and man is made patent through the well-known route of appealing to the ambivalent character of a black cat with almost human qualities. We also find in the novel a temptation to play with some areas of the occult. The beautiful Sopholite evokes the Councillor's daughter; a Theopolite in the crystal ball reminds us of student Anselmus (and of all unfulfilled poets) in the ninth Vigil of *The Golden Flower Pot*.[53] The symbolic salamander makes also a final, brief and dramatic appearance in Nic Nac's story, in the fireworks finale of Sophopolis.

There are also deep differences: for Hoffmann, as for Frederick Seele, the two parallel worlds are irreconcilable and, furthermore, there is nothing one can do to connect them. For Holmberg it is essential to build bridges, to link or unite them, otherwise there will be no progress in Argentine society or in science there.

However, literary influences on Holmberg are by no means exclusively confined to the fantastic. The impact of Charles Dickens, the only literary author specifically quoted in *Nic Nac*,[54] must not be forgotten when discussing Holmberg's sharp critique of social life in Aurelia. Dickens helped

---

53 *The Golden Flower Pot (Der goldne Topf)* first appeared in E.T.A. Hofffmann, *Fantasiestücke in Callots Manier* (1814).

54 Holmberg (1876), p. 116n.

him to articulate a coherent criticism of both social relations and social conditions, particularly in the capital city, where the author had gained a more direct personal experience. Holmberg's admiration for the work of Dickens ran deep: later, he translated some of Dickens's works into Spanish.

Finally, his journey, or dream, threw Nic Nac into a lunatic asylum. As in Hoffmann's fiction, the strong emotions of the trip seem to have tipped the balance of his mind and resulted in Dr Uriarte's diagnosing him as suffering from *Planetary Mania*. Now, he could either cling to his visions or, like Galileo, deny the reality of his journey, which had been the most important event in his life. He had been properly initiated, and took the first and most gracious option, ending at the mercy of showers in a sort of cold inferno, to have his thoughts cooled down and washed out. The figure of the asylum may itself be a reflection on the complex and multifaceted Argentine society of the time.

Although a powerful, real story can be read behind the fantasies of *Nic Nac*, this novel represents an important point in the history of literature in Argentina, when fantasy begins to break away from realism and constitute itself as a separate genre.

5

Literature and Science in Martínez Estrada's Work

Alejandro Kaufman

I

The relationship between science and literature has been an object of consideration, study and debate since the emergence of modern science. It forms the background to this study of the writings of Martínez Estrada on culture. Its history can be traced back to the tensions existing at the beginning of Enlightenment. The period from the 1950s up to the end of the twentieth century have been marked by this debate. The refocusing of anthropological studies towards the modern city as an endogenous phenomenon seen as strange contributed to the emergence of a new notion of culture. The techno-scientific revolution which provided a link from the birth of cybernetics, molecular biology and the sciences of the mind to the construction of artificial languages signalled some of the milestones of a vast and heterogeneous field that brought into question the relationship between science and culture. In addition to this, there is C.P. Snow's positing of the 'two cultures'[1] and the many debates that it engendered. I shall refer to two of them in particular because they serve as examples of the actuality of the topic as a reflection on cultural analysis, beyond the specific relevance of the arguments exposed in that polemic or whether it is thought to be of only historic value.

Wolf Lepenies claims that in his study of the relationship between science, literature and sociology he attempts to describe

> ... the contention between two groups of intellectuals: on one hand the men of letters, i.e. the writers and critics, on the other the social scientists, above all the sociologists. From the middle of the nineteenth century onwards literature and sociology contested with one another the claim to offer the key orientation for modern civilisation and to constitute the guide to living appropriate to industrial society. This contention

---

1   C.P. Snow (1964) *The Two Cultures and a Second Look* (Cambridge).

played a significant role in the public life firstly of France and England, then also of Germany: its consequences are still visible today.[2]

The second half of the twentieth century[3] has seen the shattering of any illusion of conceptual unity regarding a coherent and comprehensive account of contemporary knowledge. The collapse of a unifying framework is linked to the quantitative growth of knowledge and to the expansion of discursive productions, even in the field of the creative imagination. This growth has made it impossible for an individual mind to attain a cognitive or evaluative judgement of the available knowledge. However, the diversity and heterogeneity of the conceptual frames is verifiable, resulting not only in an enormous increase in ideas within the different disciplines but also in some cases in alternative paradigms within the different disciplines and fields of knowledge.[4] Initiatives like those described by Brockman appear in this same context, but with a different meaning from the one assigned by Lepenies:

> The third culture consists of those scientists and other thinkers in the empirical world who, through their work and expository writing, are taking the place of the traditional intellectual in rendering visible the deeper meanings of our lives, redefining who and what we are.[5]

The changes that took place in the latter half century brought consequences: 'A 1950s education in Freud, Marx and modernism is not a sufficient qualification for a thinking person in the 1990s'. Thus, for a better definition we must refer to the 'two cultures' discussion. Related to Snow's book, Brockman says:

> On the one hand, there were the literary intellectuals; on the other, the scientists. He noted with incredulity that during the 1930s the literary intellectuals, while no one was looking, took to referring to themselves as 'the intellectuals', as though there were no others. This new defini-

2   Wolf Lepenies (1985) *Die Drei Kulturen* (Munich), English translation: *Between Literature and Science: The Rise of Sociology*, (New York, Paris, 1988), p.1.

3   Periodisations depend on historical distinctions as well as those between disciplines. The quotation from Lepenies' refers to the polemic between positivism and hermeneutics. The reference to the last half century remits to the changes of paradigm that have taken place since then, reformulating actual modes of previous discussions.

4   Of course, this is more valid in humanities and social sciences than in the so-called 'hard' sciences.

5   John Brokman (1995) *The Third Culture* (New York). An anthology of discussions including, among others, authors like Stephen Jay Gould, Richard Dawkins, Francisco Varela and Daniel Dennett.

tion by the 'men of letters' excluded scientists such as the astronomer Edwin Hubble, the mathematician John von Neumann, the cyberneticist Norbert Wiener, and the physicists Albert Einstein, Niels Bohr, and Werner Heisenberg.[6]

The expression 'third culture' had been used by Snow in the second edition of his book,[7]

> ... in which he optimistically suggested that a new culture, a 'third culture', would emerge and close the communications gap between the literary intellectuals' world and in Snow's third culture, the literary intellectuals would be on speaking terms with the scientists. Although I borrow Snow's phrase, it does not describe the third culture he predicted. Literary intellectuals are not communicating with scientists. Scientists are communicating directly with the public. Traditional intellectual media played a vertical game: journalists wrote up and professors wrote down. Today, third-culture thinkers tend to avoid the middleman and endeavour to express their deepest thoughts in a manner accessible to the intelligent reading public.

Further on he affirms that what traditionally had been called 'science' was then known as 'public culture'.[8] He then introduces the main ideas discussed in the anthology:

> The ideas presented are speculative; they represent the frontiers of knowledge in areas of evolutionary biology, genetics, computer science, neurophysiology, psychology and physics. Some of the fundamental questions posed are: Where did the universe come from? Where did life come from? Where did the mind come from? Emerging out of the third culture is a new natural philosophy, founded on the realisation of the import of complexity, of evolution. Very complex systems — whether organisms, brains, the biosphere or the universe itself — were not constructed by design; all have evolved. There is a new set of metaphors to describe ourselves, our minds, the universe and all of the things we know in it, and it is the intellectuals with these new ideas and images — those scientists doing things and writing their own books — who drive our times.[9]

---

6   *Ibid.*
7   Quoted in note 1.
8   Brokman (1995), p. 18.
9   *Ibid.*, p. 21.

The heterogeneous nature of the transformations that have taken place, as well as their speed and magnitude, explains in part the distance that separates the different authors and trends that somehow characterise this situation from those others who remain or attempt to remain within the boundaries of their knowledge, and of what they can develop as cognitive creation. One of the consequences of this extraordinary magnitude of expansion is the emphasis or even the exclusivity which imprints itself more and more on formal stylistic or procedural aspects.[10] These acquire a definitive role in facilitating the development of a common language in those areas where a common language has become impossible.

II

When reading Ezequiel Martínez Estrada's texts on culture, one possible risk is the misunderstanding produced by two categorial axes of great signification in cultural analysis. The first consists of trying to reduce to known disciplines and fields of study (literature, fiction, sociology) discourses that deliberately exceed those boundaries. The other is rooted in a problem that lies at the heart of what could be considered 'a definition of culture'. This faces expositive and institutional difficulties related to the transformations that have taken place only recently, since the second post-war period, namely, the matter of values. The debate about values has been undermined by the conjunction between the ethos of science — in its paradigmatic form — and the methodological and formal shift in institutions related to knowledge. This debate has not simply been 'repressed', as sometimes happens, for complementary phenomena are taking place. The performative dimension of cognitive processes implies values that are not perceptible or expressible other than from outside of a previously established framework. Also, at a subjective level, notions such as criticism, negativity and pessimism have become commonplaces that lead to reductive interpretations and implicit interdictions.[11]

10 Some of Brockman expressions and attitudes concerning 'traditional intellectuals' have an appearance of exclusion rather than interpellation.

11 Andrés Avellaneda describes it as it follows: '*Radiografía de la pampa*, en 1933, y *La cabeza de Goliat*, en 1940, inauguran su fama de "nihilista", juicio que aún hoy se repite con ligereza. Su método de vastas generalizaciones e inducciones; su búsqueda de mitos y arquetipos para explicar la realidad; su subordinación de las categorías históricas a diseños ético-sociales, producen esos textos en que la visión 'negativa' de la realidad nacional termina por ofender el gusto de tirios y troyanos. Desde entonces, pesimista es el adjetivo que se asocia a su nombre, como una muletilla.' 'Martínez Estrada, revolucionario,' *Cuadernos Americanos*, vol. CCLXIV, Segunda Época, no. 1, Jan–Feb. 1986, p. 159.

These interdictions consist of an impulsive, thoughtless exercise of value judgement that attributes negativity or pessimism to a determined thought, changing it for a 'constructive' citadel. This is why misunderstanding takes place: because discourses so constructed are not always capable of such reductions without altering their meanings. When they are being read without those interdictions, they cannot be conceived outside of the available framework of actions and discourses other than in conflictive interaction with them.

For this perspective, the critical exercise does not mean an elaboration from the outside of what constitutes the object of the analysis but rather an exposition of the internal forces or tensions of it. This means the admission of contradiction as inherent to the object. Not as an anomaly but as the drive that underlies future transformations. On confronting heterogeneous discourses the instrument of cultural analysis does not result in a series of categorially linked statements. It produces a heteroclite series of juxtaposed formulations, coexisting in their tensions, and of unpredictable results. Such unpredictability and internal opposition differentiates this kind of discourse from the conventional scientific procedures which obey existing paradigms. Therefore, the 'negativity' of cultural analysis is part of the analysed processes and is articulated with the existent analysis in a 'positive' sense. Sometimes, Ezequiel Martínez Estrada expresses himself in terms of *expectancy* regarding the progressive and synergetic quality of science and technology within liberal societies. Therefore, he states:

> *La situación que los gobiernos dictatoriales plantean a los pueblos, en cuanto se relaciona con la cultura popular, es muy complicada. Por una parte restringen la libertad de adquirir y difundir cierto tipo de cultura que entienden desfavorable a sus designios políticos, y por otra infunden coactivamente la vigencia obligatoria de un tipo de cultura dirigida que cohonesta esos designios. Todavía no se ha legislado la implantación de un tipo de cultura de fuste correlativa a los adelantos del orbe industrializado, es cierto; pero de hecho se han ido colocando al margen de las formas de pensar consentidas por los regímenes de fuerza.*[12]

The neglect of density in Martínez Estrada's thought results in the exclusion of those of his statements that are transversal to usual disciplinary and axiological axes. To point this out in terms of degrees of density and hermeneutic crossings does not imply the indiscriminate acceptance of all discourses that ask to be included in this characterisation. There is not, and this is precisely the point, a procedure which standardises judgements in

---

12   Ezequiel Martínez Estrada (1967) *Análisis funcional de la cultura* (Buenos Aires), p. 91.

specific cases. According to J.F. Lyotard,[13] the obstacle experienced by the critico-philosophical thought does not originate so much in the *matter* as in the *quantified time* that cases demand. Given that time has become a scarce commodity, it supposes an implicit value. Therefore, the demand of quantified time required by a particular discourse will become in itself a reason for its exclusion. According to this, the magnitude and speed which characterises the exponential growth of our knowledge are determining factors for the inclusion or exclusion of a discourse.

III

The conceptual outline underlying the present chapter postulates a reading frame which brings to light that which cannot be perceived in the work of Martínez Estrada when it is considered as 'literature' in the accepted sense of the term. A 'current' reading of Martínez Estrada's texts in dialogue with the present cultural field reveals unexpected meanings. For Martínez Estrada the essay form is simply a means of cultural and political intervention. In him we do not find an author who accepts the distinction between literature and science as discrete disciplines. The metaphorical use of scientific language to describe socio-historical facts, as systematised in the critical edition of *Radiografía de la pampa*,[14] is not a free and stylistic employment of scientific terminology. If we consider Martínez Estrada 'seriously', we will find in his familiarity with such terminology a complex coexistence with scientific language, to which he refers allusively. Therefore, his writing becomes an alternative to that of those thinkers who accept implicitly their separateness from the world of science and thus curtail the range of their thought.

The critical project of Martínez Estrada finds a synthetic formulation in the last paragraph of *Radiografía de la pampa*. Its conceptual development is continued, nearly 30 years later, in *Análisis funcional de la cultura*.[15] It is

13   J.F. Lyotard (1983) *Le Différend* (Minuit: Paris).

14   Ezequiel Martínez Estrada (1991) *Radiografía de la pampa*, Critical edition coordinated by Leo Pollman (Madrid: Consejo Superior de Investigaciones Científicas).

15   Ezequiel Martínez Estrada wrote his essay *Análisis funcional de la cultura* in 1960. It was published in Buenos Aires by Centro Editor de América Latina in 1967. It is a theoretical text that includes many references to various disciplines and constitutes a significant intervention by Martínez Estrada in the debate on culture. His study is also relevant because of its contemporaneity with discussions that took place in other parts of the world. Its repercussion has grown since its publication.

worth quoting the above-mentioned paragraph in full to illustrate the total and comprehensive construction which Martínez Estrada sought:

> *Lo que Sarmiento no vio es que civilización y barbarie eran una misma cosa, como fuerzas centrífugas y centrípetas de un sistema en equilibrio. No vio que la ciudad era como el campo y que dentro de los cuerpos nuevos reencarnaban las almas de los muertos. Esa barbarie vencida, todos aquellos vicios y fallas de estructuración y de contenido, habían tomado el aspecto de la verdad, de la prosperidad, de los adelantos mecánicos y culturales. Los baluartes de la civilización habían sido invadidos por espectros que se creían aniquilados, y todo un mundo sometido a los hábitos y normas de la civilización eran los nuevos aspectos de lo cierto y de lo irremisible. Conforme esa obra y esa vida inmensas van cayendo en el olvido, vuelve a nosotros la realidad profunda. Tenemos que aceptarla con valor, para que deje de perturbarnos; traerla a la conciencia, para que se esfume y podamos vivir unidos en la salud.[16]*

A critical thought which does not postulate itself merely in opposition to that which has been previously established, but as a pointer to its internal contradictions is, in turn, understood as a socio-historical thought and not as a set of abstract statements to be substituted by other abstract statements. The conclusion belongs to the lineage of stoicism. The question of knowledge does not refer to the origin of the world, of life or of the human mind. It does not exclude them or leave them aside but it formulates another question: How is it possible to live in a world where death, pain, illness and poverty exist permanently? Modern science constitutes another lineage; that of eudaemonia, or the Promethean endeavour of transforming the world and overcoming temporality. These interrogations are in conflict with each other and though they exist in different dimensions, yet they intersect.

A tentative list containing the terms of scientific origin which denote Estrada's attitude towards science, indicates what he thinks about the public use of science. The shifts that he brings to bear on the meanings of scientific terms suggests that, for him, they belong to the public linguistic heritage.[17]

16 Martínez Estrada (1991), p. 256.
17 That tentative list taken from *Radiografía de la pampa's* critic edition (pp. 263–315) suggests the following words: alotropía, alotrópico, anastomosarse, anhidro, apófisis, asintótico, brasílide, cámbrico-silúrico, cariocinesis, catabólico, catódico, cenozoica, coloidal, córvidos, cretáceo, criptógamo, cristalográfico, devónico, diastasa, didelfo, digitígrado, dimerosómato, dióptrico, diplópico, eclíptico, eoceno, epizootia, exosmosis, ferroprusiato, filogénesis, fisiparidad, galvánico, gástrula, geosinclinal, gimnospermo, gliptodonte, glosolalia, gondwánide, gulfstream, galófilo, hidrópico, intususcepción, jurásico, loess, megalocefálico, mendeliano, milodonte, nefelococcigia, oligoceno, onfálico, paleozoico, parénquima, pinnípedo, placentario, pleistoceno, polípero, proboscídeo, seudópodo, siluriano, superfetación, superfetado, tectónico, tendinoso, teratológico, tremedal, trófico, wolfram.

Discussion about the employment of the language of science outside its frontiers becomes a discussion about public liberties, no matter how much it may be disguised by erudite 'correctness' or epistemological rigor. In a liberal context those actions objectively considered to be dangerous to others are the only ones that can justify legitimate restrictions.[18]

Readings which underline the literary quality of Martínez Estrada's thought miss the political and cognitive dimension that runs through it. In Martínez Estrada, the aesthetic is articulated with the axiological:

> *El hombre libre económicamente no es tan peligroso para un status de injusticia como el hombre intelectualmente libre. Si en algún país se obtuviera la liberación económica del hombre sin su liberación intelectual y ética, la especie entera permanecería sometida, esclavizada.*[19]

Social criticism relates to the problem of freedom, overcoming economism. The 'scientist' in Marx gets subordinate to the 'philosopher', a thinker concerned with freedom and justice:

> *... si la injusticia social existe no es solo por la diferencia de clases económicas, sino también por la diferencia de grados de cultura y de conciencia del bien y del mal en el sentido socrático. Somos injustos porque somos ignorantes y malvados, y ésta no es cuestión económica, como Marx lo sabía perfectamente bien y sus prosélitos lo olvidan. Precisamente esto es marxismo: crear la conciencia de la esclavitud del hombre a sus fantasmas, que son sus peores amos, mucho más que a los patrones y al capitalismo. El*

---

18 To conform with Sokal's 'political correctness', Martínez Estrada should be incorporated in to the list of corrupters whom he denounces in his book *Impostures intellectuelles*. Sokal does not inquire about the context in which terms and concepts are employed. He considers that public employment of knowledge must be subordinated to an expert opinion. He does not think that speculative productions of thinkers and men of letters that take place outside the scientific domain, therefore not affecting that specific domain, can show us some relevant way to goals that are, again, extraneous to specific scientific matters but absolutely appropriate within their domain. Insofar as he ignores this distinction, his provocation becomes harmless for those authors whom he discusses, though it is detrimental to obtain an amplified view of public liberties in their intellectual aspect. Sokal's spectrum of what is acceptable is quite narrow. He does not appeal to demonstrable but to virtual consequences — that is to say intellectual ones — and in that sense, his attitude can be considered authoritarian. Neither Szasz nor Feyerabend accept those restrictions when they become justified in the attempt to prevent 'demonstrable' consequences, and they extend the public sphere to boundaries that cannot be imagined by Sokal because of his limitations. Martínez Estrada's position is evidently rather more moderate than that of the radical authors mentioned above.

19 Martínez Estrada (1967), p.82

hombre es el esclavizador de sí mismo porque no sabe lo que hace, ni le duele, y para esto lo educan. El marxismo mal entendido, llamado 'marxismo vulgar' por los teóricos, no se detiene a considerar que la emancipación de las clases trabajadoras no significa apenas algo deseable si no se le emancipa de la concepción teocrática, metafísica y al par utilitaria de la vida que las mantiene sojuzgadas. Pues están sojuzgadas porque no pueden razonar sensata, rectamente, es decir, con libertad. Para aliviar esta penuria, Marx, que era humanista antes que economista, imaginó una complicada teoría del capitalismo y de toda la Economía Política. El verdadero marxismo quiere liberar al hombre de sí tanto como del amo, porque él suele ser su amo peor, el 'amo de sí mismo' que lo somete y expolia. El fin del marxismo es la libertad, y para obtener la libertad humana proclama la necesidad de la libertad económica.[20]

Freedom, regarded as a goal, requires special consideration as to the mediations necessary to achieve it: 'La cultura para el pueblo no puede ir contra estos fines sagrados'.[21] Martínez Estrada embarks upon a cultural criticism in terms that cannot but be aesthetic but which are fundamentally motivated by a politically libertarian disquiet: 'para que el trabajador permanezca al pie de la máquina existen complicadísimos mecanismos de periódicos, radios, literatura, arte, espectáculo y hasta ciencias aplicadas y recreativas. Educar mal al pueblo es someterlo.' For Martínez Estrada, mass culture constitutes a system of 'servitude within liberty or, as expressed by Max Weber, 'slavery without masters'.[22]

IV

Martínez Estrada remains cautious in his opposition to critical positions relating to science, and adopts balanced counterpositions when he does not act cautiously. The first attitude can be illustrated by the following:

... si se considera la civilización occidental como un status, la cultura puede aún descender a grados muchísimo más bajos sin detrimento para el funcionamiento regular de su maquinaria. Hasta es posible que lo que hemos entendido hasta hace poco por cultura carezca de aplicación útil en una clase de organización de la vida social, cuyo automatismo solo requiere un tipo de conocimientos prácticos de medición, extraños a una axiología propiamente dicha.[23]

20   *Ibid.*, pp. 82–3.
21   *Ibid.*, p. 83.
22   *Ibid.*, p. 25.
23   *Ibid.*, p. 92.

What is relevant in this assertion, beyond its obvious opposition to the established, is that the way in which it is formulated denotes an anxiety for a more integrated, harmonious, peaceful and fair society. The deep structural discrepancies, which critical thought uncovers, are a gloomy threat on the horizon.

Martínez Estrada's depressive attitude has been pointed out: '[el carácter] atrabiliario de algunos conceptos que Martínez Estrada vierte sobre la ciencia[24] cuando debe compararla o contraponerla con las humanidades en general'.[25]

> *Siempre permanecerá en esta actitud. Los avances tecnológicos nunca parecieron alegrarlo sino más bien lo contrario. Dijo en su Discurso en la Universidad:*[26] *'Yo sé que desgraciadamente hay que aceptar las atrocidades de una civilización industrializada y mecanizada, pero no quiero que sirváis a esos menesteres que pueden dirigir muy bien los ingenieros de fábrica, los agrónomos y los veterinarios. Ellos no creen en mis ciencias del espíritu, que quizás consideren entre las ciencias ocultas, y por réplica yo no creo en sus ciencias manuales. Si tuviera que seguir dialogando con los númenes máximos de la tecnología, Ford, Taylor o Molotov, en vez que con Montaigne, Thoreau y Nietzche, me sentiría muy desdichado. Por lo cual debo replicar a quienes me dicen que no entrarán nunca en mis templos de los ídolos, que yo nunca entraré en sus ferreterías.*[27]

In this case the problem of 'literariness' does not arise because we are dealing with a 'political' text. Martínez Estrada is stating a truth: his discourse is that of a freethinker who speaks as an individual. We cannot find in him signs of misanthropy or ill humour, only an adherence to the particular norms of freethinking as laid out by a long, ancient tradition. Such personal formulations are the preconditions for the necessary time and concentration for an examination of diverging aspects of a complex and painful reality. Free thought follows a *via dolorosa* in search of salvation. For Martínez Estrada, such salvation may even prove to be cathartic.

---

24   The confusion between science and technology becomes obvious in our references. Although it is true that Martínez Estrada refers to technology rather than to science when considering knowledge, the example is valid for the problems of reading analysed in this chapter. On the one hand, the complex and even controversial relationship between science, technology and politics, and, on the other one, the oblique readings of what is quoted as 'pessimism'.

25   Nidia L. Burgos (1996) 'Martínez Estrada y la universidad,' *Cuadernos Americanos*, vol. 4, año X, Nueva época, no. 58 (July–Aug.,UNAM, México), p. 106.

26   *Ibid.* The quotation is from Ezequiel Martínez Estrada (1959) *Discurso en la Universidad* (Bahía Blanca: Universidad Nacional del Sur), pp. 17–8.

27   Burgos (1996), p. 106.

To return to the counterposition mentioned earlier, this is supported by the following quotation from the same source, where our author considers the deep transformations he proposes for education:

> ... *mi parecer es que cuanto se intente y se lleve a cabo (y esto sería lo peor) para renovar, modificar y perfeccionar el status actual de la enseñanza sería tarea inútil si no se ajusta a una concepción científica, racional, desechando todo prejuicio de naturaleza extraña a la enseñanza misma. Puedo asegurar que de antiguo, y todavía hoy, con los problemas educacionales se imbrican otros extraños. A mi modo de ver, la enseñanza se relaciona con el saber y no con la fe. ... Un plan educacional como yo lo concibo configura una estructuración filosófica más que técnica, como lo exigirá el tratamiento ulterior de este plan en su posible aplicación a la realidad.*[28]

In conclusion, a reading of Martínez Estrada not only as a literarily figure but also as an important thinker situates him in his rightful place in any contemporary discussion on culture, to be considered alongside such figures as T.S. Eliot and George Steiner as one of the leading ponderers of the definition of culture.

---

28   Ibid., p. 110. Estrada's quotation is from 'Anteproyecto de reestructuración de la enseñanza,' first part, *El Atlántico*, 25 May 1956. Emphasis added.

6

The *Nature* Effect in Latin American Science Publications:
The Case of the Journal *Redes*[*]

Claudio Canaparo

> But what a scientist — or indeed any reflective person —
> categorises as his or her environment and then causally re-
> lates to observed behaviour, is always a part of that ob-
> server's domain of experience and not an independent ex-
> ternal world.[1]

Redes in the World

The 'línea editorial'

The series of editorials ('editoriales') that appear at the front of each issue of the journal constitute the most substantial source for the editorial line ('línea editorial') of the journal.[2] A second key state-ment did not appear in the journal, but was informally circulated by the staff under the form of a supplement in September of 1998.[3]

The editorial line conveys the general direction of thought and ideas that the journal expresses as determinant in relation to its *discourse*, that is, in relation to the journal's perspective *in toto* and not necessarily related to the writings as individual products. In other words, it is the *ground* over which the discourse of the journal is going to be precipitated since it is this discourse which tries to situate the journal in a broader context — either in relation to other publications, authors or elements. This is what is meant by *Redes* in the *world* ('mundo').

*   This is a part of a work in progress on: *The 'Nature Effect' in Latin American Science Publications*, and of a larger project on the analysis of the relationship between science and writing to be published.

1   Ernst von Glasersfeld (1995) Radical Constructivism (London: Falmer Press), p. 15.

2   See, for example, issue no. 6, p. 6.

3   See, *Redes* (1998) 'Informe de Septiembre de 1998', memo published informally by the editors of the magazine.

The editorial line also presents the paradox which was to characterise the relationship between its enunciates ('enunciados') and its discourse. According to the editors, the scientificity ('cientificidad') that the journal tries to analyse must be situated, in a nineteenth-century fashion, in the context of each individual country.

> *REDES, como publicación y como proyecto, es la invitación a una nueva mirada colectiva sobre la realidad de nuestros países y la significación que para ellos tienen la ciencia y la tecnología.*[4]

Similar comments can regularly be found throughout the 15 issues considered.[5]

Furthermore — and probably to avoid any association with the popular ideological perspectives of the 1960s[6]— the journal establishes that:

> *No tenemos la pretensión de generar un pensamineto latinoamericano, en el sentido de algún modelo teórico que se arrogara el propósito de iluminar rumbos, pero sí queremos ejercitar el pensamiento, tomando conciencia de que la incertidumbre debe ser asumida y que en las brumas prospectivas a menudo sólo nos quedan, como elementos orientadores, la capacidad de comprender nuestras limitaciones y oportunidades, y el sentido ético que nos impulsa al logro de una sociedad mejor.*[7]

However, this 'pensamiento latinoamericano' is immediately and paradoxically reformulated:

> *REDES y quienes la hacemos no pretendemos comunicar nuestras certezas sino nuestras incertezas; creemos que es hora más de formular preguntas que de ofrecer respuestas; queremos someter a crítica muchos de los supuestos básicos de las políticas en ciencia y tecnología de los países de la región; queremos, en resumen, estimular 'un pensamiento latinoamericano', rescatando la pluralidad de enfoques sin renunciar a la búsqueda del consenso sobre las oposiciones y dilemas básicos en esta materia.*[8]

4   'REDES, as a publication and as a project, is the invitation to a new collective look to the reality of our countries and the meaning that science and technology have within them.' Issue no. 8, p. 5.

5   See, for example, issue no. 9, p. 7; no. 10, p. 5; no. 15, p. 5 and ff.

6   See, for example, F.H. Cardoso and E. Faletto (1969), *Dependencia y desarrollo en América Latina* (Mexico City: Siglo XXI).

7   'We do not intend to generate a Latin American thought, not in the sense of a theoretical model which proposes to illuminate the way to follow, but we would certainly like to exercise the thought, while aware that the uncertain should be assumed and that sometimes only the capacity to understand our limitations and opportunities remains together with the ethical sense that encourages us to build a better society.' Issue no. 8, p. 6.

8   'REDES and the people that make it do not intend to communicate the certainties but the uncertainties. We believe that it is not the time to offer answers but to for

The realisation of a 'pensamiento latinoamericano' seems here to coexist — if one follows the privileged placing granted by the journal to European authors (and theories) of high visibility (Latour, Collins, Bourdieu, etc.) — with the aim of diffusion of the situation of European academics in the area ('región').[9] However, this apparently open intellectual attitude to uncertainties and answers — that is assumed to be natural to the 'pensamiento latinoamericano'— does not always manifest itself in the 'editoriales' of the journal.[10]

Redes — given the explicit editorial statement that the journal has the purpose of creating 'un proyecto editorial destinado a los estudios sociales de la ciencia y la tecnología en Latinoamérica'.[11]— could indeed be included inside a general market and academic phenomenon that has emerged over the last 30 years as 'Social Studies of Science' or 'Social Studies of Knowledge'.[12] A specific contextualisation of the argument is offered by the 'Director' (editor) of the journal.[13]

Under these conditions — and in contradiction of the editor's comments — the journal did not present or develop any original insights. The fact that privileged space was allocated to authors like K. Knorr Cetina[14] or to B. Latour[15] is highly revealing. However, more than the presence of these authors or the analysis dedicated to them, it is the absence of any reflection as to how, where and why these authors have come to constitute an *inevitable bibliography* in Latin American academic environments — especially if one considers that those environments manifest their interest in a 'pensamiento latinoamericano'.

It is evident that the 'línea editorial' of the journal is trying to place the journal — in terms of historiography, symbolism and references— in the sphere of 'Social Studies of Science' of the Anglo Saxon world (*world* in its broader expression):

---

mulate questions. We want to analyse and challenge the basis of the science and technology policies in the countries of the region. We want to stimulate a 'pensamiento latinoamericano', rescuing the plurality of perspectives without resigning the search for a consensus within the basic disputes in the matter.' Issue no. 8, p. 6.

9   See, for example, Pablo Kreimer (1998) 'REDES and the Building of a Latin American Tradition in STS Studies,' in *EASST Review*, vol. 17, no. 4.

10 See, for example, issue no. 15.

11 'A publishing project aimed at social studies of science and technology in Latin America,' issue no. 10, p. 5.

12 See also issue no. 2, p. 4; no. 3, p. 7; no. 4, pp. 5 and 7; no. 6, p. 6.

13 See issue no. 2, pp. 5–8.

14 See issue no. 7, p. 6.

15 See issue no. 9, p. 7.

REDES se articula en una tradición relativamente cercana en el tiempo que considera la ciencia como una actividad social producida históricamente.[16]

It is also possible to conclude that the British journal *Social Studies of Science* is the model that the editor has in mind.[17] One of the few articles regarding the situation of scientific publications and science studies in Latin America, published by a high profile journal, leads to the same conclusion.[18] Book reviews and commentaries related to these authors, which are frequently published in *Social Studies of Science*, aim to consolidate this aspiration of the editor.[19]

In the *Informe* (report) distributed off the record by the journal, the editors put forward a perspective of 'estudios sociales de la ciencia' (social studies of science) by which, from a historiographical generalisation, they try to anchor the journal in the Anglo Saxon tradition.[20] However, at the same time, there are references to a 'pensamiento latinoamericano en ciencia, tecnología y sociedad' and assurances that the journal

Sostiene la convicción de que, como en otros campos de la cultura, es posible y necesario desarrollar una identidad propia de los países latinoamericanos en el campo del conocimiento científico y que a través de ella es como se podrá lograr una inserción más provechosa para nuestras sociedades, en el contexto de la ciencia mundial.[21]

This contrast will be more evident at the moment of confrontation between the 'enunciados' and the 'discurso' of the journal. On the other hand, this situation becomes more relevant when one realises that the 'estudios sociales de la ciencia' became the leitmotif and ultimate aim of the 'cientificidad' (scientificity) employed by the journal.

16   '*REDES* is articulated in a relatively new tradition which considers science as a social activity produced in a historical way.' *Redes* (1998), p. 2.
17   See, for example, Kreimer (1998).
18   See Hebe M. Vessuri (1987a) 'The Social Study of Science in Latin America,' in *Social Studies of Science*, vol. 17, pp. 519–54. It is no surprise that this article was published in *Social Studies of Science* as a 'regional report', while, for example, none of the other articles related to the same topic received the indication of 'regional' (the example of the French case is probably the most clear). The reasons are evident.
19   See, for example, issues no. 7, no. 9 and no. 13)
20   *Redes* (1998), p. 3; and also see Pablo Kreimer (1994) 'Estudios sociales de la ciencia: algunos aspectos de la conformación de un campo,' in *Redes*, vol. 1, no. 2, pp. 77–105 and Kreimer (1998), *op. cit.*
21   'Sustains the conviction that, as in other cultural fields, it is possible and necessary to establish an autonomous identity of Latin American countries in the area of scientific knowledge, and that, through it, a better insertion may be obtained in the context of world science.' *Redes* (1998), p. 2, and also issue no. 1, p. 6.

In this way, the idea of science that the journal proposes — by associating scientific development with political compartmentalisation by countries or 'regiones' (regions) — has a clear nineteenth-century European style.[22] Thus, the editor's quotation, in the first issue, of B. Houssay (1887–1971) and the association made between cultural and national identity development and the constitution of a 'ciencia propia' (our own science).[23] This stance was maintained by the journal throughout the whole collection and would enable the merging of science, technology and politics that characterises a high proportion of articles.

The coexistence between these two worlds, one immediate and local ('pensamiento latinoamericano') and another European and global ('Social Studies of Science', etc.) constitutes a consequence or a theoretical version of the coexistence of two opposite perspectives at the level of the 'enunciados' and at the level of the 'discurso'. The fact that Amílcar Herrera and Bruno Latour get together in the pages of the journal is not something that can be explained through the historiography but rather via the format of the journal, which aimed to put together the epistemic presuppositions discussed in the area of 'Social Studies of Science' and the specific needs that the editors identified in the 'sociedad latinoamericana' and in its academic, publishing and university spheres.

Institutionalisation

Different to the so-called *established* publications, and more akin to other Latin American publications that try to function in partnership with academic institutions (mostly universities), *Redes* does not function autonomously and try to use its partnership to consolidate its position in the market.[24] The association of the journal with the Universidad de Quilmes and with the 'Instituto de Estudios Sociales de la Ciencia y la Tecnología'[25] should be analysed in this light — adding to the fact that, right from the first issue, *Redes* appeared as a kind of 'órgano' for the projection of that

---

22   See for example *Nature*, vol. 121, pp. 525–6.

23   Issue no. 1, p. 6. For an analysis exploring the form in which the Nobel Prizewinner B. Houssay established an paradigm of science in Argentina see Alfonso Buch (2000a) 'Forma y función de un sujeto moderno. Bernardo Houssay y la fisiología argentina (1900–1943),' PhD thesis, 170 pp.

24   Many cases, in various disciplines, could be mentioned regarding this situation. Among others: *Interciencia* (Venezuela), *Revista da Sociedade Brasileira de História da Ciencia* (Brazil), *Confines* (Argentina), *Margem* (Sao Paulo), *Historia y Grafía* (México).

25   See issue no. 8, p. 5.

university.[26] The phantom of *Social Studies of Science* — established in 1971 in association with the 'Science Unit' of the University of Edinburgh — is also evident here.

It is necessary to note that the Universidad de Quilmes also acts as the official publisher of the journal, which means that this partnership is more than just a formal agreement.[27] The extent to which the journal would be able to survive without the support of the university is highlighted by the problems over the publication of the journal after issue number 15 (August 2000).

Even if the aim of the editors to gain a wide audience has been stated,[28] the exhibition of academic *signals* ('señales') has always functioned in the journal as a way of indicating seriousness and academic rigor. The constant mention of the *academic place* of the authors is, for example, one of these elements. The journal *must* be an academic publication,[29] and if, at the same time, it is directed to a relatively wide audience this constitutes a bonus. This is the reason why there is reference to a 'reflexión académica' (academic reflection) for various audiences.[30]

The Local Market

Together with the re-establishment of the non-military system of government in Argentina at the beginning of the 1980s — a phenomenon which characterised a great number of countries in Latin America — there emerged a multiple series of cultural and institutional initiatives.[31]

The introduction of an Anglo Saxon idea of 'new journalism' constitutes an innovative element in the publishing market of the River Plate area.[32] It is evident that this 'new journalism' was adapted and came to generate a *sui generis* version and, contrary to what happened with its Anglo Saxon counterpart, affected different areas and disciplines in universities. If the magazine *Humor*—published from 1978 — could be considered the first example *avant-la-lettre*, the journal *Pagina 12* — a sort of 'rioplatense' *The Guardian* — its purest expression.

---

26   See issue no. 3, p. 8; no. 4, p. 5; no. 6, p. 7; no. 9, p. 5; and issue no. 10, p. 7.
27   Issue no. 1, cover and p. 3.
28   See, for example, issue no. 10, p. 6.
29   See, for example, issue no. 11, p. 5.
30   Issue no. 11, p. 5.
31   See, for example, Kaufman (1998).
32   See for example Osvaldo Soriano (1987) *Rebeldes, soñadores y fugitivos* (Buenos Aires: Editora/12).

Figure 6.1: Relationship between the Thematic Division and the Design of the Journal as Regards the Canonical Form of *Nature* with the Sources Employed by *Redes*.

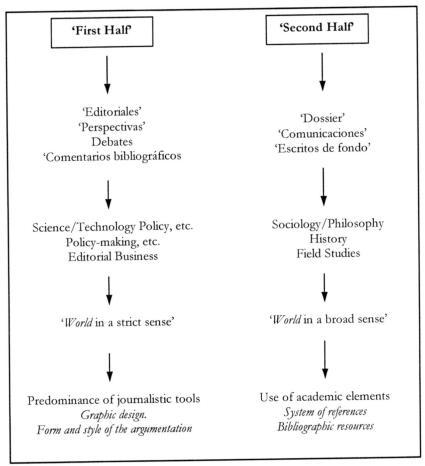

It is in this context of renewal of the discursive and enunciative forms in the *media* — but not in the established organism of scientific diffusion — that *Redes* appeared for the first time in September 1994. The journal joins a series of academic publications that could be categorised together. Journals with different origins — like *Nueva Sociedad, Ciencia Hoy, Punto de Vista, Confines* and *El Ojo Mocho* — have, however, a common factor whereby, as in the case of *Redes*, their 'Cultural Studies', 'New Journalism' and 'Science Studies' coexist under the same 'línea editorial'.

This brief historiographical consideration is relevant because *Redes* oscillates between the deployment of academic elements (references, bibliography) and the use of journalistic tools (forms of argumentation, graphic design) (see Figure 6.1).

If, on the one hand, the content is academic, on the other hand, the discourse of the journal conveys a form of journalistic construction and vulgarisation ('difusión').

The World and the Population

From the observation of *Redes*'s collection — in relation to authors, their academic affiliations, their status within the journal and the type of writing they produced — it is possible to establish some conclusions which have been presented within the generic map of authors published by the journal (see Table 6.3). For analytical purposes — and as they appear in the journal— the journal's members have been divided into staff in a strict sense ('administrative staff') and members of the 'Consejo Asesor' ('academic staff'). However, this distinction disappears from issue number 15, in which all the members — excluding the 'technicians'— are part of the 'Consejo Editorial'.[33]

**Figure 6.2: Map of the Authors, Members of the 'Comité Asesor'
and Members of Staff.**

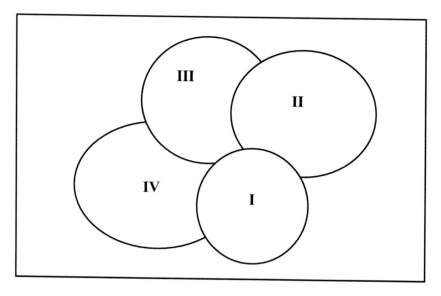

33	Issue no. 15, p. 2.

I = circle of staff and members of the 'Consejo Asesor' (*journal's circle*).

II = circle of authors who belong to the institution which publishes the journal (*journal's institution circle*).

III = circle of authors related to the institutions to which the staff or members of the 'Consejo Asesor' belong (*members' institution circle*).

IV = circle of authors without apparent direct institutional relation (*authors' circle without established institutional relation*).

If the authors published by the journal are considered in terms of their writings and in relation to the institutions to which they belong, they could be divided into four *circles*, according to the division made in Figure 6.2 and in Tables 6.3 and 6.4a. The idea of overlapping circles is perhaps the most accurate to represent the situation of the authors regarding the journal.

If group I is the core of authors that make up the journal — it is important to note that all the declared members of staff *have published* in the journal, that is, they are all authors — the rest of *circles* then could be defined in relation to the distance that separates them from this core in the following manner:

Under this description the journal could be conceived as a group of authors gathered together around a *cuaternaria* hierarchy and in relation to a *discurso rector* (principal discourse, editorial line). The participation of the majority of the staff not only in the administration of the journal but also as edited authors is maybe one of the most common characteristics of Latin American publications (science and social science) with a low IF and with structural problems of distribution.[34]

With a total of 177 authors considered (one piece published = one author), circle I (59 authors) represents 33.3 per cent of the total, while circle II represents 5.6 per cent. Regarding this last group it should be noted

---

34   See, for example, Kaufman (1998).

that, except for the articles, the rest of the content (mostly book reviews) shows no indication of any affiliations of the authors. It is very likely that many authors who produce book reviews ('reseñas') are part of this circle II, but the information available is insufficient for this conclusion (see Table 6.4a). Thus, the relatively high proportion of the total represented by circle IV (23.7 per cent). Finally, circle III with 42 authors represents 32.2 per cent of the total.

38.9 per cent of the authors edited by the journal within its six years of existence is composed either by members of the journal or by authors who have a close relationship with the institution that edits the journal. If one adds to this figure the 32.2 per cent of authors related indirectly through other academic institutions to the journal, then the proportion increases to 71.1 per cent. In other words, *Redes* is a journal which publishes the same authors that made it or authors that have a direct link with the institutional circles related to these first mentioned authors.

Of a total of 63 authors of book reviews, 22.2 per cent of them are by authors belonging to circle I, 11.1 per cent to circle II and 30.8 per cent to circle IV. However, it was not possible to identify any author related with circle III (see Table 6.4a). Members of the journal (staff and 'Consejo Asesor') and authors connected to the institutions to which the first are also connected constitute 33.3 per cent of book reviewers edited. Furthermore, 28.5 per cent of the 63 book reviews edited are the product of authors who have published in other sections of the journal and who are part of circle I, II or III (see Table 6.4b). Therefore, taking all the figures into account, one could say that 61.8 per cent of book reviews are by authors who have a direct relationship with the journal.

If one considers this situation in geographical terms a clearer picture of the distribution of the population of *Redes* emerges. Figure 6.3 offers the basic elements of this possible map showing that the population is concentrated around Buenos Aires and in particular around members of the journal and institutions to which they belong.[35]

Of a total of 159 authors studied in the period 1994–2000, 58 of them were members of the journal staff (36.4 per cent), eight were from the institution that edits the journal (5 per cent), 20 were related to the University of Buenos Aires (12.5 per cent) and 27 were from various in-

---

35   The affiliation considered is that offered by the journal and only one author has been considered by article. In cases of many authors, every author has been considered separately and in cases where many authors are part of the same institution these have been considered as one author only (see also Tables 6.4a and 6.4b).

stitutions of Buenos Aires, its environs and in some cases from other states ('provincias') (16.9 per cent, which include some non identified cases).

Figure 6.3: Population of the Journal with its Geographical Location

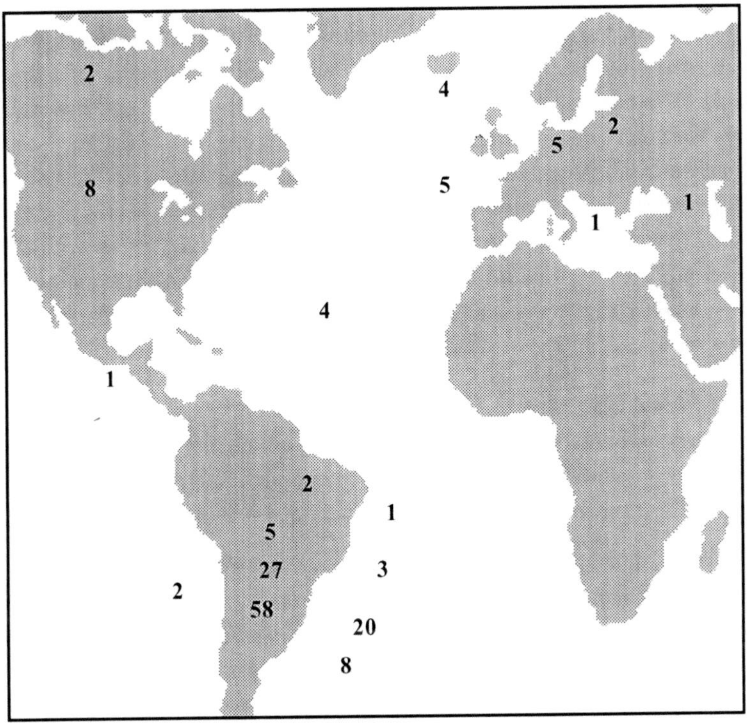

Considering this same information, the following distribution could be established: two authors belong to Chilean institutions (1.2 per cent); five are from Campinas; two from Sao Paulo; one from Rio de Janeiro (Brazil: 5 per cent); four from Caracas (2.5 per cent) and one from Costa Rica (0.6 per cent). The rest of the population could be distributed as follows: two authors in Canada (1.2 per cent); eight in the United States (5 per cent); four in the UK (2.5 per cent); five in France (3.1 per cent); five in Spain (3.1 per cent); one in Italy (0.6 per cent); and two in the rest of Europe (1.2 per cent).

It is interesting to note that the journal positioned its discourse — especially in a theoretical sense— around a core composed by the minority of authors coming from the USA and Europe (16.7 per cent). If one compares this distribution of population with the origin of the language of the writings (Table 6.1a) the correspondence is high. But if one observes that the discourse of the journal bases its visibility in the 22 per cent of translated literature (Table 6.1b) and that the references, quotes and footnotes deployed in the articles direct the reader to a literature written in English or French, then the map presents a paradoxical situation since it is a minority of non Latin American authors who shoulder the conceptual parameters and bases of the sociological/philosophical analysis of the journal (see Table 6.2) —analysis and concepts which in the end are the elements that give identity to the journal.[36]

Finally, if one compares this map with the distribution of the subscription declared by the editors, the preceding observations seem to be confirmed (Table 6.5). Subscriptions are concentrated around circles I, II and III (Figure 6.2), that is, around those institutions to which authors belong. The same editors publicly recognised that the distribution of the journal constitutes a basic problem that they had been unable to overcome.[37]

Readers, Market and Culture

The innovative and renewal character with which the editors present the journal[38] could not be excised from the publishing market to which the journal affirmed to be orientated. According to the editors,

> REDES no ha sido, a lo largo de sus diez números, solamente un proyecto editorial dirigido a la comunidad académica. Por el contrario, se trató, y se trata, de un proyecto cultural, entendido en un sentido amplio: debatir en la escena pública los aspectos de la ciencia y la tecnología que interesan a nuestras sociedades. Es en este sentido que ha sido y sigue siendo fundamental la interlocución con otros sujetos sociales: los científicos y tecnólogos, en primer lugar. Ellos se han acercado a REDES de un modo creciente, mostrando sus preocupaciones, ofreciendo su perspectiva, sus intereses. En segundo lugar, los investigadores en ciencias sociales, quienes se van mostrando cada vez más interesados por los estudios sobre la ciencia y la tecnología, y sus relaciones con otros objetos del conocimiento social. En tercer lugar, los tomadores de decisión en América Latina, quienes encontraron, a lo largo de los números de REDES, debates y críticas, información y reflexiones, en suma, la 'puesta en cuestión' del conocimiento científico y el desarrollo tecnológico y de las políticas públicas que los tienen como objeto. Finalmente, pero no por ello menos importante, diferentes actores de la sociedad se fueron interesando crecientemente acer-

36   See, for example, Kreimer (1994 and 1998).
37   *Redes* (1998), p. 4.
38   See, for example, issue no. 10, p. 6.

*ca de las investigaciones, discusiones y pensamientos que han ido expresándose en los
10 números de REDES.*[39]

This goal, in a highly hegemonic, fragmented and competitive publishing market, could be no more than a vague aspiration.[40] The 90 subscriptions to the journal constitute proof of this situation. The limited and highly concentrated reception of the journal denies the assumptions implicit in this statement, especially those referring to the non-local *ámbito*.

This aspiration to cover a wide sector of the market explains the variety of writings the journal publishes and the absence of a coherent epistemic parameter — despite the coherent graphic structure of the journal. Moreover, this variety of writings could explain the absence of a clear process regarding the assignation of scientificity to the writing and thus a clear procedure to *select* the writings.

What is deeply naïve is the belief that nowadays a 'cultural project', in academic and scientific terms, could be developed by a journal.[41] If what is understood by 'culture' refers to a *cultura letrada* (literary culture), to dedicate a journal to the analysis of science as a written phenomenon — as is the case, for example, of *Social Studies of Science* — seems a reasonable strategic decision. However, to publish a journal which consistently discusses scientific

---

39   'REDES has not only been, throughout its first 10 issues, a publishing project aimed at the academic community. On the contrary, it has been and still is a cultural project understood in a broad sense to debate in the public sphere aspects of science and technology that are of interest to our societies. It is in this sense that the interaction with other social subjects (mainly scientists and technologists) has been and still is fundamental. It is these scientists and technologists who have approached *Redes* in increasing numbers, showing their concerns and offering their perspectives and interests. Secondly, there are researchers in social sciences who show an increasing interest in studies related to science and technology and their links with other social subjects. Thirdly, decision-makers in Latin America who find throughout Redes debates and critiques, information and thoughts, that is, the questioning of scientific knowledge and technological development, as well as the public policies regarding them. Last but not least, different actors of the society have shown an increasing interest regarding in the research, debates and thoughts that have been present in the 10 issues of *Redes*.' Issue no. 10, p. 6.

40   See for example Yves Jeanneret (1994) *Écrire la science. Formes et enjeux de la vulgarisation* (Paris: PUF): Massimiano Bucchi (2000) *La scienza in pubblico. Percorsi nella comunicazione scientifica* (Milan: McGraw-Hill); Bernadette Bensaude-Vincent (2000) *L'opinion publique et la science* (Paris: Institut d'Édition); Bertha Gutiérrez Rodilla (1998) *La ciencia empieza en la palabra* (Barcelona: Península).

41   See, for example, Eduardo Ortiz (2000) 'Studies through Four Books by Rowe and Canaparo,' unpublished review, pp. 1-8.

policies, as with *Redes*, does not seem a common sense decision if what is wanted is to approach the above-mentioned idea of culture.

On the contrary, if the notion of 'culture' is referred to in a broader sense — not only to the *letter* ('letra') but to other forms of collective manifestation — then the situation is even less comprehensible from the point of view of the journal since a notion of visually and cinematographically orientated 'culture' — like the present one[42] — can hardly be affected or modified by a publication started in 1994 and which, four years later, possessed only 1,500 readers[43] in a local market that, according to non-official estimations, is populated by 100,000 readers.[44]

Moreover, the belief that a journal can be a privileged form of delivering a project focused in a 'pensamiento latinoamericano',[45] perforce ignores institutional and academic practices (at least as arguments of analysis) which constitute relevant epistemic elements in the erection of the above-mentioned 'pensamineto latinoamericano'.[46]

Nature-*ism in the World*

The idea of a limited perspective (local world) and a wider one (global world), in which the journal tries to find a place, is based fundamentally in a *Nature*-ist perspective of scientific production ('scientificity') (see Figure 6.1):

> *La ciencia y la tecnología fueron pensadas en buena medida, desde Bacon en adelante, como vías para el dominio y control de la naturaleza. Sin embargo, la maestría alcanzada en el dominio de la naturaleza lejos está de encontrar un reflejo equivalente en los asuntos humanos, especialmente en América Latina.*[47]

This realist and human version of scientificity is the same, for example, as that which John Maddox, the former editor of *Nature*, has postulated for

---

42   See for example Daniel Jacobi (1999) *La communication scientifique. Discours, figures, modèles* (Paris: PUF) and Alexis Martinet (ed.) (1994) *Le cinéma et la science* (Paris: CNRS Édition).

43   Redes (1998), p. 9.

44   See, for example, Alejandro Agostinelli (2000) Comments about scientific publications in Argentina, e-mail correspondence.

45   See for example issue no. 3, pp. 5–8; no. 6, pp. 5–7 and no. 8, pp. 5–6.

46   See, for example, Jeanneret (1994).

47   'Science and technology were thought to be, from Bacon onwards, ways of dominating and controlling nature. Nevertheless, the mastery reached with regard to the domination of nature is far from finding an equivalent in human matters, especially in Latin America.' Issue no. 13, p. 5.

years through the pages of the journal. The existence of a 'real world' associated with an idea of nature, with science as a tool to face that reality and scientific accumulation ('progress') as a way of making a democratic society constitute the three fundamental elements of what has been indicated as *Nature*-ism[48] and which the 'línea editorial' of *Redes* expressed here very clearly.[49] This perspective was also presented by the editor of the journal as 'una utopía "razonable"' (a reasonable utopia)[50] in the sense that it endeavours to be fitting to the local requirements.[51]

A natural consequence of this *Nature*-ism is the specificity that has been attributed to the 'sociedades latinoamericanas' (Latin American societies).[52] In other words, the editorial perspective ('línea editorial'), after being inscribed in a general context (global world), tries to identify a particular identity or specificity (local world) from the initial parameters.[53]

Without specific reference to any general context (global world), the 'línea editorial' has nevertheless introduced in its scientific perspective an epistemic notion of *Nature*-ism which works more or less under these conditions. This notion could be indicated by saying that what *the journal discuss are the administrative applications and formulations of science and not the epistemological quality or status of this form of scientific production.*

The Scientific Administration

The constant mention made to institutional events and the insistence with which the journal (editorial line mostly) is associated with the organisation of committees, congresses and associations[54] could indeed be interpreted not only as a strategy oriented towards the participation in a political debate regarding scientific policies (government administration of science) but also as a sort of syndrome coming from the intellectual dependence with which the editorial line decided to adopt authors with high *visibility* in the market (Bourdieu, Latour, etc.).

---

48   See Claudio Canaparo (1999) *El efecto* Nature (London: La Protesta Ediciones) and by the same author (2004) *Ciencia y escritura* (Buenos Aires: Zibaldone Editores).
49   See also, for example, issue no. 11, p. 5 and no. 13, p. 6.
50   Issue no. 6, p. 6.
51   Issue no. 2, p. 8.
52   Issue no. 13, p. 5.
53   Issue no. 13, pp. 5–7 and no. 2, p. 8.
54   See, for example, issue no. 2, p. 4; no. 4, p. 5; no. 7, p. 5; no. 8, p. 5; and no. 9, p. 5.

This situation could also explain why there is a confusion — in the editorial line of the journal — between field studies of scientific communities and epistemological debates regarding how scientificity is constructed.[55] Michael Polanyi pointed out, more than 30 years ago, this *way of escape* where talking (writing, making an exposition, etc.) about a particular subject is confused (misplaced) with the possible reflection on how and why that subject has a function in the scientific and academic community.[56]

The few writings of the journal that try to deal with institutional aspects which could lead into an epistemological debate unfortunately finished, in the context of the journal's discourse, mostly either confirming the *Nature*-ism of the publisher and scientific market[57] or appealing to a 'voluntarismo expresivo' (hope expression) based on historical re-formulations of theories or academic perspectives ('modernity', 'Frankfurt' orientated or economical development approaches) in tune with the editorial line of the journal.[58]

Finally, within the same dichotomy come the mentions to 'Social Studies of Science'[59] and to those elements related to administrative aspects of scientific production.[60]

---

55   See, for example, issue no. 8, p. 7.
56   See Michael Polanyi (1973) *Personal Knowledge. Towards a Post-Critical Philosophy* (London: Routledge and Kegan Paul, first edition 1958).
57   See, for example, issue no. 2, pp. 41–76; no. 2, pp. 77–105; no. 6, pp. 13–32; no. 7, pp. 161–92; no. 8, pp. 11–46; no. 12, pp. 51–73; no. 10, pp. 155–78; no. 11, pp. 231–55; no. 12, pp. 51–79; no. 13, pp. 13–48; no. 15, pp. 15–44.
58   See, for example, issue no. 1, pp. 9–26; no. 1, pp. 73–112; no. 2, pp. 5–26; no. 6, pp. 33–76; no. 11, pp. 37–52; and no. 12, pp. 111–7.
59   See, for example, issue no. 2, p. 4; no. 4, p. 7; no. 6, p. 6; no. 10, p. 5; and no. 14, p. 5.
60   See, for example, issue no. 3, pp. 13–58; no. 4, pp. 97–128; no. 7, pp. 13–51; no. 10, pp. 13–58; no. 10, pp. 115–78; no. 15, pp. 5–10.

Table 6.1a: Comparative table in relation with the origin (language) of the writings edited by the journal. It is a partial account but it clearly shows the tendency of the journal. Book reviews are not included but the writings without identifiable author are included.

Issue	Amount of writings considered	Literature in Spanish-American	Literature in Portuguese	Literature in English	Literature in French	Literature in German	Literature in Spanish
Number 1	12	10	1		1		
Number 2	13	11			2		
Number 3	7	6		1			
Number 4	8	7			1		
Number 5	14	11	2	1			
Number 6	8	5		1	1		1
Number 7	10	7	1	1	1		
Number 8	11	8	1	2			
Number 9	9	7		1			1
Number 10	9	9					
Number 11	12	9	1		1		1
Number 12	9	8		1			
Number 13	8	5	2	1			
Number 14	10	5	2	2	1		
Number 15	9	5	1	1	1		
Special Number	5	1	2				3
	154	114	13	12	9	0	6

Table 6.1b

Issue	Amount of writings considered	Translated literature
Number 1	12	2
Number 2	13	2
Number 3	7	1
Number 4	8	1
Number 5	14	3
Number 6	8	2
Number 7	10	3
Number 8	11	3
Number 9	9	1
Number 10	9	
Number 11	12	2
Number 12	9	1
Number 13	8	3
Number 14	10	5
Number 15	9	1
Special Number	5	4
	154	**34**

Table 6.2: Generic classification of the writings according to their content. Articles written by more than one author are considered only under one of them. Book reviews are not included and only the writings that appear in the index of each issue have been considered.

Issue	Amount of articles	Sociology, Philosophy (group 1)	History (group 2)	Science/technology Policy Public Understanding of Science (group 3)	Field Studies (group 4)	Policy making, Government Policies (group 5)	Editorial business ('linea editorial') (group 6)
Number 1	12	3	1	7			1
Number 2	13	5	3	1		3	1
Number 3	7	2	2	2		2	1
Number 4	8			2		3	1
Number 5	14	3	5	3	1	1	1
Number 6	8	3	1	1	1	1	1
Number 7	10	3	1	4		1	1
Number 8	11	2	1	7			1
Number 9	9	2	3	2	1		1
Number 10	10	1	2	6		1	1
Number 11	13	3	2	5	1	1	1
Number 12	11	2		5		3	1
Number 13	9	1	1	4		2	1
Number 14	13	1	2	7	1	1	1
Number 15	9	1		1	1	5	1
Special Number	5			1		3	1
	162	31	24	58	6	27	16

Table 6.3

Cartographic placement of the journal's writings according to the author's journal's status and institutional situation. The 'Special Number' and book reviews are not included. The placement is made according to four parameters: (1) authors members of the journal, (2) authors with relation with the institution which edit the journal, (3) authors affiliated to one of the institutions to which the journal's members belong, (4) authors without direct relation known.

Issue	Total authors	Journal's circle	Journal's institution circle	Members' institutions circle	Rest of authors
Number 1	12	7		3	1
Number 2	13	7	2	4	1
Number 3	7	2	1	2	2
Number 4	9	5		3	
Number 5	14	5	1	6	2
Number 6	11	2		4	5
Number 7	14	5		6	1
Number 8	13	2		4	5
Number 9	10	3	1	4	2
Number 10	14	4	1	5	2
Number 11	16	6	1	7	2
Number 12	10	3	2	2	3
Number 13	11	2	1	4	3
Number 14	12	4		2	6
Number 15	11	3		1	7

Table 6.4

Cartographic placement of book reviews ('reseñas') according to the relation that authors have in relation to the journal. The 'Special Number' is not considered. The parameters are the same used in Table 6.3.

Table 6.4b offers a comparison among the authors that, on top of at least one book review, have been also writing in other sections of the journal.

Table 6.4a:

Issue	Total reviews	Journal's circle (members)	Members' institutional circle	Unknown
Number 1	6	1		2
Number 2	5	2		1
Number 3	4	1	1	1
Number 4	4	1	3	1
Number 5	4	1	1	
Number 6	5	2	1	1
Number 7	4			2
Number 8	3	1		1
Number 9	4			3
Number 10	3	2		1
Number 11	4			3
Number 12	5	2		2
Number 13	4	1		1
Number 14	4			2
Number 15	4		1	3
	63	14	7	24

Table 6.4b

Issue	Total reviews	Authors of articles
Number 1	6	3
Number 2	5	2
Number 3	4	2
Number 4	4	1
Number 5	4	
Number 6	5	1
Number 7	4	1
Number 8	3	1
Number 9	4	1
Number 10	3	1
Number 11	4	1
Number 12	5	1
Number 13	4	2
Number 14	4	2
Number 15	4	
	63	**19**

Table 6.5

Distribution of the journal's subscriptions according to official estimations of *Redes* (1998, p. 8).

Countries and areas	Subscriptions
Argentina	67
Latin America	16
Europe	4
United States	3
Total	**90**

7

Two Scientific Traditions in *Martín Fierro*

Norma S. Horenstein

> I thought, however, that judging the past from
> the perspective of one's own present-day views
> is the worst sin that any historian can commit.
> David Hull[1]

Introduction

Western science recognises, though not exclusively of course, two main scientific traditions or, following Kuhn, two dominant paradigms in the exact sciences: the Pythagorean in mathematics and the Newtonian in physics. The level of consolidation of these paradigms can be measured according to their permanence: more than a thousand years for the former and several centuries for the latter. Newtonian conceptions dominated the field of physics until the emergence of Einstein's theories. On the other hand, discussion in the field of mathematics is an endless controversy. Even nowadays both mathematicians and philosophers of mathematics assume either Platonist (realist) positions or, by contrast, anti-Platonist, anti-realistic or nominalist views. Yet, neither of these approaches has been able completely to disqualify the other. Both traditions have proponents and followers that afford good reasons to support either realism or nominalism. In other words, we can find strong arguments by means of which Platonists try to prove that universals like numbers, functions, etc. are just discovered by men. But there are equally strong arguments to assert that we live in a world of individuals, that universals are simply men's creation, *ex nihilo* creation.

In this contribution I aim to show that in the verses of the *payada*, the author of Martín Fierro — José Hernández — supports a conception of numbers, weights and measures that must be interpreted as Pythagorean–Platonist on its mathematical side and Newtonian on what is relevant to its physics.

---

1   D. Hull. (2001) 'The Great Success of a "Foul Book",' *Science*, vol. 291, p. 833.

Martín Fierro

José Hernández published the first part of *Martín Fierro* in Buenos Aires at the end of 1872. The success of this edition led the author to publish a second part, which was entitled *The Return of Martín Fierro*. Although Hernández in this second part of the poem tries to make thorough use of the popularity of his hero and intends to show the opposition of the gauchos to corrupt officialdom, the texts we are interested in are those of the *contrapunto* or *payada* (meaning a singing-match), where part of the call and response imply some reference to science. Those verses are to be found in the *Thirtieth Canto* of the second part of the poem.

José Hernández was born in 1834 in an estancia that was the property of the patrician family of his mother. Biographers of Hernández claim that during his childhood he accompanied his father in the rough tasks of their farm; that he did not have the opportunity of receiving a formal education; and that he had no tutors during his childhood. The education of children of traditional wealthy families was usually in the hands of tutors and governesses, but this, apparently, was not the case in Hernández's family. At the age of 19 José Hernández joined the army, but in 1861 he left it and started his career as a journalist and a politician.

Considering Hernández's lack of formal education, perhaps it is problematic to attempt to identify in his main work, *Martín Fierro*, the influences of the Pythagorean–Platonic as well as Newtonian traditions. Ezequiel Martínez Estrada has pointed out that

> Hernández's greatness emerges from his lack of culture and even of the preoccupation of acquiring it, from his life devoted to action and from his journalist work. There are no reasons to admit that he was the kind of man fond of reading or a person worried either about what we now understand as technical knowledge obtained in books, or in methodical study. A man of the nineteenth century, that of our Illustration ... he remained indifferent to the eagerness of knowledge and perfection characteristic of the men of his generation and class.[2]

All that has been written about what he used to read must — in the words of Martínez Estrada — [should] be rejected as an attempt to attribute to him a knowledge he neither had nor needed;[3] he did not even have a li-

---

2   Ezequiel Martínez Estrada (1948) *Muerte y Transfiguración de Martín Fierro* (Mexico City: FCE), p. 41.

3   *Ibid.*, p. 42.

brary (only Avellaneda recalls he had one). According to this interpretation, Pagés Larraya maintains that 'to confer Hernández a wisdom he did not have is to concoct a picture of Hernández that conforms to the academics' taste'.[4]

However, in *El mito gaucho*, Carlos Astrada, avoiding a discussion on what Hernández formally studied or knew, noted the parallels between Martín Fierro's sayings and Pythagorean thoughts. He states:

> Although the enumeration of the three great units: sun, world and moon, is capricious, the reminiscence of Pythagorean triad is evident. (…). As the Pythagorean triad integrates itself and concentrates itself in the divine unity, in the gaucho's cosmogony, that triad is reabsorbed in 'the being of all beings' who, as the great monad that eternally recreates itself, forms the unity.[5]

Despite the fact that José Hernández never attended university, and that he states in a letter to the editors of *Martín Fierro* that his ideas were just the product of meditation and constant and careful observation, the verses that I shall analyse provide enough evidence to support the thesis that he had more than an elementary knowledge in the field of mathematics, physics and philosophy. For a self-educated man, it seems that Hernández, who valued poetry as a means of singing ideas, was capable of understanding and using fairly complex scientific and philosophical concepts.

By the time he wrote his poem, Hernández was almost 30 years old and had spent the previous nine years serving in the Argentine army. He may have become acquainted with the two different scientific traditions formerly mentioned (the Pythagorean–Platonic and the Newtonian) while reading or studying at the army, or perhaps before. The knowledge Hernández acquired in these fields was, certainly, profound and serious enough to allow for the extrapolation of technical matters into his poetic work. Hernández's references to his own lack of education should, perhaps, be understood as a provocative argument in relation to the high respect for the role of formal education professed by the Buenos Aires intellectual elite of the time.

When, in the *payada*, the Moreno asked Hernández about numbers, measures, weights and time, Martín Fierro was able to give an answer that offers some clues on his understanding of the universe and its origins. Martín Fierro says:

---

4   A. Pagés Larraya, (1952)*Prosas del Martín Fierro* (Buenos Aires: Ed. Raigal), p. 22.
5   Carlos Astrada, (1964) *El mito gaucho* (Buenos Aires: Ediciones Cruz del Sur), p. 96.

Come on then darkie; don't jib or shy,
In view you're so mighty wise;
You've got the hook well down, I feel,
So answer me right off the reel,
While the beat you ring on the sounding string;
What song is the song of the skies?[6]

The Pythagorean–Platonic Tradition

Let me make some brief remarks on the Pythagorean–Platonic tradition. According to Diogenes Laertius, the notions of quantity and measure were concepts that show Pythagorean inspiration, and that were introduced in Greece through Pythagoras' influence. The most fundamental Pythagorean scientific contributions are the notion of a duality between the limited and unlimited, and the equivalence of objects and numbers. Since even a musical scale depends upon the imposition of definite proportions on an indefinite continuum of sound, it seemed reasonable to think that the principles of the limited and the unlimited had universal application.

> There seems no reason to doubt the tradition that Pythagoras himself discovered — probably by measuring the appropriate lengths of string on a monochord — that the chief musical intervals are expressible in simple numerical ratios between the first four integers.[7]

Considering the fact that the whole secret of the musical scale underlies the first four integers, their sum is the number 10 (the Decade), which may embrace the whole nature of numbers according to Aristotle's interpretation. A privileged consideration of the Decade also appears in Plato's work. This last author did not invent the concepts of unit and *arithmoi*; he picked them up already developed. He was influenced by the Pythagorean doctrine, but he would not accept it as it was then understood.

---

6   [¡Ah negro! Si sos tan sabio
      No tengas ningún recelo:
        Pero has tragao el anzuelo
      Y, al compás del estrumento,
      Has de decirme al momento
      Cuál es el canto del cielo.]
      (J. Hernández, 1989), *El gaucho, Martín Fierro*, Argentina, Sainte-Claire Editora, Edición bilingüe, p. 265).
7   G.S. Kirk and J.E. Raven, (1957) *The Presocratic Philosophers* (Cambridge: Cambridge University Press), p. 229.

The meaning of the word *arithmos* in Greek literature is a 'set of units'. (In fact, the replacement of the notion of *arithmoi* by the notion of number took place much later, in the fifteenth century, as a result of further developments in mathematics.)[8] Besides, the unit is not exactly a number in the ordinary sense. The unit, being a part of the subject matter of arithmetic, is not an *arithmos* because the latter is a plurality of units, and hence more than one of them. Plato did not believe that the series of mathematical numbers is the result of mathematical operations.[9] On the contrary, arithmetical operations cannot be defined on *arithmoi* because they are not natural numbers. Mathematical numbers exist by participation in ideal archetypes. In Philebus 15b, Plato asks:

> ... how, if each of them [the Ideas] is eternally self-identical ... it can yet be confidently pronounced to be the unit it is; finally must we take it to have suffered dispersion and become many in the indefinite multiplicity of the realm of becoming, or can it be as a whole outside itself, and thus come to be one and identical in one thing and in several at once — a view which might be thought the most impossible of all?.[10]

The common Greek definition of number, as 'plurality of units', informs us of what numbers are for, not what numbers really are. In this sense Plato's theory of number is the necessary complement of the Pythagorean conception of number held in Greece before his time. The weakness of Pythagorean doctrine of number derives from the confusion between sensible and intelligible levels. Aristotelian criticism of Pythagoreans emphasises the difficulty in conceiving material realities and spiritual ones both as number. According to Kirk and Raven:

> The unfortunate consequence of their diagrammatic representation of numbers was that the Pythagoreans, thinking of numbers as spatially extended and confusing the point of geometry with the unit of arithmetic, tended to imagine both alike as possessing magnitude.[11]

Plato tried to avoid these difficulties by developing a more complex ontological picture. In Plato's mature version of the *Theory of Ideas* we find

---

8   However, the notion of *arithmoi* must not be identified with the notion of natural number. Besides, the modern conception of number includes more than just natural numbers. For instance, negative integers, rational and real numbers. The previous remarks explain why Plato did not think *arithmoi* as possible subject matter of *arithêtikê*.

9   See, for instance, Phaedo 101 ab.

10   Philebus 15b, Plato (1956) *Philebus and Epinomis* (London: Thomas Nelson and Sons), translation A. E. Taylor, p. 108.

11   Kirk and Raven (1957), p. 246 f.

the distinction between ideal numbers, mathematical numbers and numerable sensible things. Mathematical numbers differ from sensible objects in that they are eternal and immutable, and from ideal numbers because each idea is unique, while mathematical numbers are multiple and repeatable. These two characteristics of mathematical numbers constitute the condition that makes it possible to carry out arithmetical operations. Aristotle thought that, insofar as Plato considered that the One is essence and that the ideal numbers cause things, he spoke as a Pythagorean, even though the latter did not recognise the existence of the intermediate entities, i.e., mathematical numbers. Aristotle ascribes to Plato an ontology made up of three stages, that distinguish the intermediate mathematical objects from perceptible objects and forms (archetypes).

It is worth noting the resemblance between what Martín Fierro says about the nature of number and Plato's claim in the *Epinomis*, 976d–e. Plato says:

> Let us then first consider what single science there is, of all those we have, such that were it removed from mankind, or had it never made its appearance, man would become the most thoughtless and foolish of creatures. (…) For, if we, so to say, take one science with another, 'tis that which has given our kind the knowledge of number that would affect us thus', and I believe I may say that 'tis not so much our luck as a god who preserves us by his gift of it'.

For Plato mathematical and ideal numbers must be kept separated from sensible countable things and, consequently, ideal and mathematical numbers can be considered a gift given by God to mankind, keeping men responsible for the development and application of these notions to what furnishes the universe. A similar idea is to be found in Hernández's verses:

> The sun is one, and the world is one,
> One moon in the sky we see;
> So it's plain and broad, that Almighty God
> Never made any quantity.
> The One of all ones is a single whole,
> and One was the first amount;
> Number only began to be made by man
> As soon as he learned to count.[12]

---

12 *Uno es el sol, uno el mundo,*
    Sola y única es la luna;
    Ansí, han de saber que Dios

From Hernández' point of view — as for Pythagoras and Plato — the unit is not a number, and mathematical operations, such as counting, for instance, can be done only with numbers. In *Martín Fierro*, Hernández simplifies Plato's ontology reducing it from three to two stages: we do not find in his poem the Platonian distinction between mathematical numbers and numerable things, both appear subsumed under the same category.

In order to complete the analysis of the elements of the Pythagorean–Platonic tradition present in the *Payada* we must consider the characterisation of the notion of measure given by Hernández. Measure, like number, is a notion that historians of philosophy and science attribute to Pythagoras. The verses in which Martín Fierro offers the definition of measure are the following:

> Hark well to me — if you don't agree,
> Let my ignorance be excuse —
> Every single measure man measures with,
> Man made for his private use.
> It's easy to see God didn't need
> Any measure to help his plan,
> He had nothing to measure, once he'd fixed
> The length of the life of Man.[13]

In Philolaus, a member of the Pythagorean or Italian School, we find a conception of measure opposite to that of Protagoras, who thought that man is the measure of everything. In the Pythagorean tradition God measures man's life; men by means of their souls participate in the reason of the Unity and this is the view supported by Philolaus under whom the

---

No crió cantidad ninguna.
El ser de todos los seres
Sólo formó la unidá;
Lo demás lo ha criado el hombre
Después que aprendió a contar.

13   *Escuchá con atención*
   Lo que en mi ignorancia arguyo:
   La medida la inventó
   El hombre para bien suyo.
   Y la razón no te asombre,
   Pues es fácil presumir:
   Dios no tenía que medir
   Sino la vida del hombre.
   (Hernández, 1989, p. 276).

Pythagorean tradition flourished. In the sixth fragment Philolaus affirms that the essence of things, which is eternal, admits divine knowledge but not human knowledge. The opposition between divine and human science is typical of this conception and is clearly exposed both in connection with the elucidation of the idea of number as well as with reference to the notion of measure. The ideas of measure and number were synthesised by Pythagoreans in the concept of harmony: the movement of stars produces a harmony. The speeds of the movements of the sun, the moon and all the stars measured by their distances are in the same ratio as musical concordances; the sound given forth by the circular movement of the celestial bodies is a harmony or, as Hernández would say, 'a song of the skies'.

The Newtonian Tradition

The origin of number and measure is in God; the same can be said about weight. Regarding weight, Martín Fierro answers the question the Moreno put to him in the following terms:

> God keeps in the stores of His secret lore
> This mystery profound;
> He simply ordained that every weight
> Should fall till it hit the ground.
> And since life is a bundle of bad and good,
> I'll answer you this again:
> The use of weight is to estimate
> The sins of the sons of men.[14]

Carlos Astrada (1964) points out that in these verses weight is interpreted in a double sense, with their respective scientific and historic signs, as gravitation — Newton's apple — and as Sin — Adam and Eve's apple, which determined the loss of all paradises and the beginning of history.[15] In this context we are interested in the first of these two meanings of the term 'weight'.

14   *Dios guarda entre sus secretos*
      El secreto que eso encierra
      Y mandó que todo peso
      Cayera siempre a la tierra;
      Y sigún compriendo yo,
      Dende que hay bienes y males
      Fue el peso para pesar
      Las culpas de los mortales.
      (Hernández, 1989, p. 277).
15   See Carlos Astrada (1964), p. 97.

Weight is the result of the actions gravity exerts on each of the molecules of a body. Gravity is the quality by which any body tends to the centre of earth and gravitation is the effect of universal attraction among bodies and origin of weight. These three interconnected concepts appear in Martín Fierro's answer (quoted above) to the third part of the question posited by the Negro singer.

> I'll ask you now that I'd like to know
> Since to answer me is your pleasure;
> I'll give you best in our singing-match
> If you answer this batch with all dispatch:
> Explain to me please what Number is,
> And Time, and Weight and Measure.[16]

Kant, when discussing the different kinds of judgements — in *Prolegomena* and in the *Critique of Pure Reason* — in order to establish the scientific character of physics, Newtonian physics, gave the following examples.

* Every body has extension.

* Every body has weight.

The first judgement is an *analytic* one, since what is asserted in the predicate is already expressed in the subject. Extension is inherent to body-ness; every body has to be thought as having extension. Contrarily, weight is not an inherent characteristic of bodies and, hence, in the judgement that expresses that 'every body has weight' the predicate adds something to the subject-concept. This is a *synthetic* judgement, the kind of judgement that broadens our knowledge because it is not tautological. We are not interested in Kantian aprioristic philosophy but it is worth noting that Kant considered Newtonian physics as the unique valid theory and in this theory weight is the result of gravitation.

Newton, in the General Scholium of his *Mathematical Principles of Natural Philosophy*, declares that the beautiful system of the sun, planets and comets could only proceed from the counsel and dominion of an intelligent and powerful being.

---

16   *Voy a hacerle mis preguntas,*
       Ya que a tanto me convida;
       Y vencerá en la partida
       Si una esplicación me da
       Sobre el tiempo y la medida,
       El peso y la cantidá.
       (Hernández, 1989, p. 275).

This Being governs all things, not as the soul of the world, but as Lord over all (…) All that diversity of natural things which we find suited to different times and places could arise from nothing but the ideas and will of a Being necessarily existing.[17]

The power of gravity explains the phenomena of the heavens, the sea and the Earth and the cause of this power proceeds from a cause that:

> … penetrates to the very centres of the sun and planets, without suffering the least diminution of its force.[18]

Bodies subject to gravitation have weight and

> God keeps in the stores of His secret lore this mystery profound.[19]

Explanation by means of gravity forms part of experimental philosophy, where particular propositions are inferred from phenomena and generalised afterwards by induction. Experiments with bodies of the same weight led Newton to state that all bodies about earth gravitate towards it, and that the weight of all are as the quantities of matter they contain, at equal distances from the centre of the earth (Second Corollary), independently from the form of the bodies. However, gravitation is not always observable.

> … since gravitation towards these bodies is to the gravitation towards the whole earth as these bodies are to the whole earth, the gravitation towards them must be far less than to fall under the observation of our senses.[20]

Gravitation may not be observable in some cases, weight may be negligible but the better we know these principles much greater is the recognition of God's hand in the organisation of the diversity of natural things and men's lives. In *Martín Fierro* one can detect a communication between science and literature, as it touches on topics of scientific traditions well established at the time Hernández wrote his poem. We must accept that Hernández could be acquainted with these scientific ideas only by means of studies carried out, if necessary, outside a traditional educational environment.

17   Newton (1952), p. 370.
18   *Ibid.*, p. 371.
19   Hernández (1989), p. 277.
20   I. Newton, (1952) *Mathematical Principles of Natural Philosophy* (Chicago: William Benton), p. 282.

8

Heisenberg's Uncertainty Principle in Contemporary Spanish American Fiction

Alicia Rivero

In *Two Cultures and the Scientific Revolution*, C.P. Snow claimed that science and literature are distinct, incompatible cultures.[1] However, studies by Hayles, Strehle, Friedman and Donley, as well as by other scholars, invalidated Snow's thesis. They proved that, despite differences, science and literature are interrelated aspects of a common culture. Both are affected by, express and shape culture.

The notion of a discontinuous universe without cause and effect at the subatomic level, unveiled by quantum physics, has been widely disseminated and forms part of our episteme. Hayles, Strehle and Friedman and Donley demonstrate that the cross-pollination of ideas in physics, philosophy and the arts (including literature) led not only to similar developments in these fields, but also to specific techniques which represent the new concepts about reality that physics ushered in during the twentieth century. As Friedman and Donley state perspicaciously:

> ... science does have a deep influence on other aspects of culture, even those that seem far removed, such as 'serious' literature. Major contemporary writers have been implicitly and explicitly affected by ideas in fundamental science, in particular by Einstein's theory of relativity and by quantum theory, even when those ideas have been filtered and distorted by the ubiquitous popularisations. Attention to these revolutions in science helps explicate and enrich understanding of literary works.[2]

The significant correlations between the findings of quantum physics and the development of twentieth-century literary strategies have been studied primarily in Anglo-European texts. Quantum physics and, particularly,

---

1   C.P. Snow (1960) *Two Cultures and the Scientific Revolution* (Cambridge: Cambridge University Press).

2   Alan J. Friedman and Carol C. Donley (1990) *Einstein as Myth and Muse* (Cambridge: Cambridge University Press), p. 3.

Heidelberg's uncertainty principle, which suggest that there is an observer-created, indeterminate reality, are equally relevant to the analysis of Spanish American contemporary texts, yet little research has been published to date on this topic.

This chapter provides an overview of how our perception of reality was profoundly altered in the twentieth century by quantum physics and the role of the new physics in some Spanish American countries. It will also probe the way in which reality is reflected in selected Spanish American texts which actively involve the reader as a participant, such as those by Borges, Cortázar, Volpi and Fuentes, but will only focus in detail on the latter's *Cristóbal Nonato*. In this polyphonic novel, Cristóbal, the principal narrator who explicitly mentions Heisenberg, affirms that observing phenomena changes them.[3] I will show that Cristóbal's statement implicates both the narrator and the reader in the creation of textual reality in a manner akin to the tenets of quantum physics, especially those espoused by Heisenberg.

Quantum Physics, the Observer and Reality

Quantum theory was developed between 1900 and 1926, when the determinism inherent in Newton's laws was shown to be inapplicable to atoms, since they 'behaved in random, uncontrollable ways which determinism could not account for', according to Pagels.[4] Hayles states that, by 1927, 'the mathematical formalism of quantum mechanics was essentially complete'.[5] Heisenberg and Bohr were two of the most influential physicists who formulated what became known as the Copenhagen Interpretation of Quantum Mechanics that replaced Newtonian mechanics.

Quantum-mechanical studies of atomic phenomena yield seemingly contradictory results, obtained from different experiments on the same object. For instance, electrons behave like waves in some experimental arrangements and like particles in others; this is called the 'wave-particle duality'. 'An electron beam passing through a single slit produces a distri-

---

3   Lynn M. Walford (1994), 'Mess as the Natural State: Echoes of Bakhtin and Heisenberg,' *Hispanic Journal*, vol. 15, no. 2, pp. 246–7, 255, briefly discusses the role of Heisenberg's observer in this novel, but relates the observer to generalised chaos and to Bahktin, without associating her/him thoroughly with quantum physics or with the role of the reader.

4   Heinz R. Pagels (1990), *The Cosmic Code: Quantum Physics as the Language of Nature* (New York: Bantam), pp. 4–5.

5   N. Katherine Hayles (1984), *The Cosmic Web: Scientific Field Models and Literary Strategies in the Twentieth Century* (Ithaca: Cornell University Press), p. 43.

bution [pattern] with most "particles" detected in line with the slit.'[6] This occurs because the electrons are behaving as if they were particles. One would expect that the design left on a screen after electrons are fired through two slits would be the same as when one slit is employed. Instead, 'experiments show that the pattern observed with both slits "open" is not the same as the pattern obtained by adding together what we see with each slit separately'.[7] This is due to interference when the electrons behave like waves. Even more surprisingly, as the astrophysicist Gribbin indicates.

> The electrons not only know whether or not both holes are open, they know whether or not we are watching them, and they adjust their behaviour accordingly. There is no clearer example of the interaction of the observer with the experiment. When we try to look at the spread-out electron wave, it collapses into a definite particle, but when we are not looking, it keeps its options open. In terms of Born's probabilities, the electron is being forced by our measurement to choose one course of action out of an array of possibilities. There is a certain probability that it could go through one hole, and an equivalent probability that it may go through the other ... The strangest thing about the standard Copenhagen interpretation of the quantum world is that it is the act of observing a system that forces it to select one of its options, which then becomes real.[8]

The anthropomorphic discourse on which scientific accounts rely[9] is evident and unavoidable in the case of counterintuitive quantum events. Gregory warns that 'Quantum mechanics is not a picture or a narrative that tells physicists *how* the outcome of an experiment comes about', in contrast to classical physics; rather, it is 'a mathematical expression that allows them to calculate *what* the outcome of an experiment will be'.[10]

Although quantum effects can be expressed more precisely mathematically,[11] paradoxes abound when one attempts to explain such experimental

6   John Gribbin (1988), *In Search of Schrödinger's Cat: Quantum Physics and Reality* (New York: Bantam), p. 164.

7   *Ibid.*, p. 167.

8   *Ibid.*, pp. 171–2.

9   For example, see Gillian Beer's analysis of the 'Problems of Description in the Language of Discovery,' in George Levine (ed.) (1987) *One Culture: Essays in Science and Literature* (Madison: University of Wisconsin Press), pp. 35–58.

10   B. Gregory (1990), *Inventing Reality: Physics as Language* (New York: John Wiley & Sons), p. 95.

11   See Neils Bohr (1963), *The Rutherford Memorial Lecture, Essays (1958–62) on Atomic Physics and Human Knowledge* (New York: Interscience), p. 60. A good introduction to the mathematical concepts used in quantum theory is by J.C. Polkinghorne (1989), *The Quantum World* (Princeton, NJ: Princeton University Press).

observations linguistically and on the basis of perceived experience. As Bohr points out with respect to relativity and quantum mechanics, we are concerned with the recognition of physical laws which lie outside the domain of our ordinary experience and which present difficulties to our accustomed forms of perception. 'We learn that these forms of perception are *idealisations* ... [I]n spite of their limitation, we can by no means dispense with those forms of perception which colour our whole language and in terms of which all experience must ultimately be expressed.'[12]

The wave-particle duality stands in contrast to classical, Newtonian physics. The latter presumed that data give a consistent picture of the object under investigation and that objects stand apart from and are unaffected by the measurements of an experimental observer, whereas quantum mechanics insists that the observer and what is under investigation form part of a related field.

Newtonian physics held that reality is composed of discrete elements which can be described precisely in space and time and that physical theories mirror the world exactly. Heisenberg's uncertainty principle shows instead that the position and velocity of atomic particles cannot be determined simultaneously. Heisenberg notes in *Physics and Philosophy* and *Physics and Beyond*,[13] as does Bohr in *Atomic Theory* and *Atomic Physics*,[14] that when observing such minuscule particles during experiments, the measuring apparatus interacts with and thereby alters the properties of the particles being observed.

In *Physics and Philosophy* Heisenberg did not view those particles as real objects which could be described unambiguously in ordinary language, but as 'potentialities' or 'possibilities'.[15] Heisenberg avers that the style of observation of scientists, who invent and manipulate measuring devices, is formed by the 'spirit of the time': culture influences their perceptions and

12   N. Bohr (1961), *Atomic Theory and the Description of Nature* (Cambridge, UK: Cambridge University Press), p. 5.

13   Werner Heisenberg (1963), *Physics and Philosophy: The Revolution in Modern Science* (London: George Allen & Unwin); *Physics and Beyond: Encounters and Conversations*, translated by Arnold J. Pomerans (New York: Harper & Row, 1971).

14   N. Bohr (1961a), *Atomic Physics and Human Knowledge* (New York: Science Editions).

15   Heisenberg (1963), pp. 154, 160. These partially bring to mind Aristotle's potentia insofar as both notions problematise cause and effect, albeit not in the same manner. Leonard Shlain (1991), *Art and Physics: Parallel Visions in Space, Time and Light* (New York: William Morrow) opines that 'In his original formulation of causality, Aristotle had allowed for the existence of an amorphous potentia between the rush of cause and the stamp of effect. It was the interface between the two where something unexpected could take place' (p. 249).

'what we perceive is not nature in itself but nature exposed to our method of questioning'.[16] Obviously, Heisenberg does not negate reality ('nature in itself', in this case), but he does question the epistemological aspects of reality to which scientists have access ('what we perceive' by means of exposing 'nature ... to our method of questioning'). He propounds that scientists do affect reality as observers during experimental arrangements. Furthermore, his theory suggests that this observer-created reality is influenced by the experimenter's *zeitgeist*.

Although Bohr concurred with Heisenberg with respect to the uncertainty relation and to the interference that instruments cause during the measuring of atomic events, he disagreed with his colleague's notion of subjectivity with respect to the observing subject. Bohr offered the concept of complementarity as an explanation of such diverse experimental results as the wave-particle duality. In his *Atomic Theory* he states that such an event can only be understood by means of 'diverse points of view that defy a unique description'.[17] He indicates that 'we are not dealing with contradictory but with complementary pictures of the phenomena, which only together offer a natural generalisation of the classical mode of description'.[18] As in the case of the wave-particle duality, the same event can only be understood by juxtaposing the two different interpretations of it in a complementary description. Because atomic phenomena can be visualised solely by means of mutually exclusive ideas, the models used to explain such events have 'just as much or, if one prefers, just as little 'reality' as the elementary particles themselves'.[19] This implies that scientists' epistemological interpretations of reality are constructs.

On the one hand, examining what constitutes reality and how we know this reality entails not only ontology and epistemology, but also cultural considerations. Latin America has hybrid multi-cultures in which the traditional coexists with the modern, as García Canclini and others have convincingly shown.[20] The process of cultural syncretism, of trans-culturation,[21] is not

16   Heisenberg (1963), pp. 96, 57.

17   Bohr (1961), p. 96.

18   *Ibid.*, p. 56.

19   *Ibid.*, p. 12.

20   Néstor García Canclini (1992), *Culturas híbridas: estrategias para entrar y salir de la modernidad* (Buenos Aires: Sudamericana).

21   The term 'transculturation' was originally utilised anthropologically by Fernando Ortiz, who coined the word. See his *Contrapunteo cubano del tabaco y el azúcar* (Caracas: Ayacucho, 1978), pp. 92–7. For the way in which the concept can be applied to literature, see, for instance, Angel Rama (1982) *Transculturación narrativa en América Latina* (Mexico, DF: Siglo Veintiuno).

new, but began during the time of the conquest of the so-called 'New World'; it continues from the colonial period to the present, with the coming into contact and mutual exchange of indigenous and foreign elements in Latin America.

Latin American authors belong to the intelligentsia who partake of the hegemonic discourse of science, which is both imported from without and is generated within Latin America itself — Latin America is not simply a 'helpless victim of a colonialist's language and image-making', but belongs to the West in a complex process of cultural construction, as González Echevarría notes incisively.[22] Besides, there have always been autochthonous elements of Latin American science, from pre-Columbian civilisations to the present day. During the period of colonial empires until modern times, Latin Americans have not merely borrowed from and imitated European or North American science in a servile manner — they have adapted and transformed western science, as well as making original contributions to it, as Saldaña, Vessuri and other scholars have demonstrated.[23]

Vessuri argues that the essence of science is the same around the world, but its style varies from country to country due to national and institutional factors that affect the organisation and work of scientists. He affirms that the social study of science calls into question the traditional notion of scientific objectivity, for 'el cuerpo de conocimiento científico aceptado en cualquier momento particular está estrechamente relacionado con el sistema social', which has ideological, economic, cultural, political and other implications.[24] The historian of science, Kuhn, and the anarchic philosopher, Feyerabend, among others, have also challenged science's claims of objectivity.[25]

22   Roberto González Echevarría (1990), *Myth and Archive: A Theory of Latin American Narrative* (Cambridge: Cambridge University Press), pp. 40–2.

23   For example, see J.J. Saldana (ed.)(1988) *Cross-Cultural Diffusion of Science: Latin America* (Mexico, DF: Sociedad Latinoamericana de Historia de las Ciencias y la Tecnología); Saldana (ed.) (1996) *Historia social de las ciencias en América Latina* (Mexico: Coordinación de Humanidades and Coordinación de la Investigación Científica, UNAM). In addition, see Hebe M.C. Vessuri (1998) 'Intercambios internacionales y estilos nacionales periféricos: aspectos de la mundialización de la ciencia,' in A. Lafuente et al. (eds.), *Mundalización de la ciencia y cultura nacional: Actas del Congreso Internacional 'Ciencia, descubrimiento y mundo colonial'* (Madrid: Doce Calles), p. 730.

24   H. Vessuri (1983) *Consideraciones, La ciencia periférica: ciencia y sociedad en Venezuela* (Caracas: Monte Avila), pp. 14–6.

25   See Thomas Kuhn (1970) *The Structure of Scientific Revolutions*, 2nd edition (Chicago: Chicago University Press) and Paul Feyerabend (1993) *Against Method*, 3rd edition (New York: Verso, 1993).

On the other hand, whose reality we are addressing beyond that of a scientific description of nature or of textual reality and just who the reader is presents difficulties in countries with varying degrees of development, access to education, different literacy rates, etc.[26] I am not begging these important questions by assuming a neo-positivist belief in the objective transparency of science, as if science were divorced from social, historical, political, economic, gender, racial, cultural and other concerns.[27] However, to address these matters adequately would require writing not one, but several books. Instead, here I will merely be able to touch upon such issues briefly when they are directly relevant to the discussion at hand.

I employ such terms as 'episteme', '*zeitgeist*' and their synonyms, being fully aware of their problematic implications, since all the cultures of a given period are not the same. Nevertheless, one can generalise about which paradigms are typical of an era without claiming that there is universal, cultural homogeneity. We will uncover how patterns that are typical of the broader cultural milieu do appear in modern Spanish American works.

A Brief Sampling of Twentieth Century Physics in Spanish America

Mexico has made significant contributions to international physics. Since the Physics Institute of the Universidad Nacional Autónoma de México was established in 1938, Gortari remarks that 'se vienen realizando investigaciones conectadas con los problemas que se consideran de mayor im-

26   Alicia Rivero [Rivero-Potter] (1991) *Autor/lector: Huidobro, Borges, Fuentes y Sarduy* (Detroit: Wayne State University Press) addressed the situation of the reader specifically in Latin America on pp. 31–2, as well the role of the reader as construct from the perspective of reader-response theory in the rest of the book. For additional information on the Latin American reader during the nineteenth and twentieth centuries, see, for example, Jean Franco (1983) *La cultura moderna en América Latina* (Mexico, DF: Grijalbo), pp. 16–8, 23–4; Rudolph Grossman (1972) *Historia y problemas de la literatura latinoamericana*, translated by Juan C. Probst (Madrid: *Revista de Occidente*), pp. 55, 228–9; García Canclini (1992) pp. 66–7, 72; Diana S. Goodrich (1986) *The Reader and the Text: Interpretative Strategies for Latin American Literatures* (Philadelphia: John Benjamins).

27   Such factors do affect the development of science and scientific research in Latin America, as elsewhere. The following scholars, among others, have shown this: Regis Cabral (1996), 'El desarollo de las ciencias exactas en América Latina y la política internacional,' in Saldana (ed.), *Historia social*, pp. 493–510; Francisco R. Sagasti (1996), *Evolución y perspectivas de la política científica y tecnológica en América Latina*, in Saldana (ed.), *Historia social*, pp. 511–33; Vessuri (1983), pp. 9–36; Vessuri (1996) *La ciencia académica en América Latina en el siglo XX*, in Saldana (ed.), *Historia social*, pp. 437–79; Vessuri (1998), pp. 725–34.

portancia en la actualidad, y los resultados son recibidos con merecida atención en el mundo científico'.[28]

In pre-Columbian times, science was an 'agente de cohesión cultural'.[29] After the positivistic period of Bolivian science, Condarco Morales points out that 'predomina la influencia de las grandes teorías modernas', such as those of Einstein, Planck and others.[30] Bolivian science was largely derivative of western science, especially in fields like physics.[31] Nevertheless, with discoveries in Bolivia pertaining to mesons and pions, Bolivian science had as much of an international impact during the twentieth century as it had in earlier periods.[32] Condarco Morales lists those as being from (1) the end of the sixteenth to the end of the seventeenth centuries in metallurgy, especially with the publication of Father Barba's monograph; (2) between the end of the eighteenth and the beginning of the nineteenth centuries with Haenke's contribution; (3) the beginning of the twentieth century and the 1970s in mining technology and mineralogy, as well as with Cárdenas's and Cevallos Tovas's studies.[33]

Einstein's visit to Argentina in 1925 is mentioned in passing by Babini as a strong stimulus for science there.[34] Ortiz fully plumbs Einsteins's role by relating it to how the modern Argentine intellectual and scientific scene developed, as well as to how Argentine science and literature were related in intellectual circles and at the level of the state.[35] Ortiz explains that

> The new mathematical physics, in which space, time and very rapid movement became central concerns, reached Argentina in the early years of this century. In 1909, a well-endowed Physics Institute was

28   Eli de Gortari (1980) 'Las matemáticas y la física,' *La ciencia en la historia de México* (Mexico, DF: Grijalbo), p. 364.

29   Ramiro Condarco Morales (1978) 'Conclusión,' *Historia del saber y la ciencia en Bolivia* (La Paz: Academia Nacional de Ciencias de Bolivia), pp. 325.

30   *Ibid.*, p. 303.

31   *Ibid.*, p. 326.

32   *Ibid.*

33   *Ibid.*, pp. 325–6.

34   José Babini (1949) 'Los estudios físicos y químicos,' *Historia de la ciencia argentina* (Mexico, DF: Fondo de Cultura Económica), p. 153.

35   Eduardo L. Ortiz (1998), 'The Transmission of Science from Europe to Argentina and its Impact in Latin America: From Lugones to Borges,' in Evelyn Fishburn (ed.), *Borges and Europe Revisited* (London: Institute of Latin American Studies, University of London), pp. 108–23. See also his 'A Convergence of Interests: Einstein's Visit to Argentina in 1925,' *Ibero-Amerikanisches Archiv*, vol. 21, nos. 1–2 (1995), 67–126.

created at the new University of La Plata, and distinguished German professors were imported to run it.[36]

The fact that European scientists played a mayor role in the development of Latin American science should not be taken to mean that science is primarily the province of the West, which was transmitted from the centre to the periphery, as was theorised by Basalla years ago. Basalla's model was popular in the past, but has proven to have limitations, and he himself has revisited his own landmark study.[37] It would be more accurate to propose that there was and still is a process of transculturation in the transmission of western science to Latin America, as was already suggested.

Martínez Sanz attests to the international legacy of modern science:

> ... a finales del siglo XIX la ciencia era ya patrimonio de toda la Humanidad. En efecto, a pesar del poderoso influjo de la ciencia alemana... la ciencia dejaba de ser peculiar o propia de una cultura o nación para tornarse internacional. A partir de aquel tiempo ya no se podrá hablar de ciencia europea o americana, porque los congresos internacionales y las publicaciones científicas ponían en contacto a los sabios y científicos de todo el mundo y, a la vez, daban a conocer en pocos meses cualquier hallazgo e investigación científica de interés, y los trabajos de unos pocos eran conocidos por la totalidad.[38]

Quantum Writing

Traditional, realistic fiction made mimetic claims that, like Newtonian physics, it represented the world without distortion, through the transparent vehicle of language. Such an erroneous stance was challenged by modern and postmodern texts, both in Spanish America and abroad.

Einstein's relativity and the Copenhagen interpretation of quantum mechanics presented a new world-view, which reconceptualised time, space and other aspects of classical physics, while metamorphosing the art, literature and philosophy of the twentieth century.[39] Friedman and Donley

---

36   Ortiz (1998), p. 111.
37   See George Basalla (1998) 'The Spread of Western Science Revisited,' in A. Lafuente et al (eds), *Mundialización de la ciencia*, pp. 599–605.
38   José L. Martínez Sanz (1992) *Relaciones científicas entre España y América* (Madrid: MAPFRE), p. 333.
39   Shlain (1991), pp. 387, 427, inverts the notion that science influences the arts. He gives conceptual priority more often to visionary, revolutionary art that foreshadows scientific advances as part of the zeitgeist of an era than vice-versa, due to his holistic perspective of the universe that the new physics promulgated.

evince that indeterminism and a lack of causality in literature became major themes and structural components of twentieth century texts as a result of quantum physics and of the philosophy that pondered the implications of the new physics.[40] They sagaciously probe how forms had to be created ... to give shape to the new ideas and the new forms in one discipline often paralleled those in others. Thus as four-dimensional geometry appeared in mathematics and physics, treating time as a dimension of space, the traditional three-dimensional representation in art expanded to include temporal dimension in avant-garde painting. The arts became concerned with time as a dimension and with space-time relationships. Physics found that there was no universal frame of reference, and multiple viewpoints showed up on paintings and in novels. Physicists indicated that the flux of a field represented the strength of its source, and artists began inventing their own 'field' forms. Relationships between objects became more important than the objects themselves. When physicists discovered they could not simultaneously find a particle's exact position and exact momentum, they admitted the basic indeterminism inherent in the quantum theory, which gives a probabilistic rather than a causal description of nature. In literature conventional consecutive plots often disappeared and standard syntactical relationships were fractured.[41]

Strehle skilfully develops these ideas. She reveals how the use of loose plots, parallel episodes, as well as temporal and spatial gaps in contemporary works are a result of the new physics; this stands in contrast to the single, continuous, causally unified plot of realism that mirrored a Newtonian universe with the omniscient, privileged perspective of the god-like author or of an absolute frame of reference.[42] Irony, polyphony, the accidental and fortuitous, in addition to open-endedness can also be employed to produce textual ambiguity and uncertainty, as Strehle evinces. The resistance of quantum literary texts to narrative closure challenges how 'conventional novels present reality as a stable and determined thing'[43] and elicits reader participation.[44]

---

40   Friedman and Donley (1990), p. 128.

41   *Ibid.*, pp. 2, 24.

42   Susan Strehle (1992) *Fiction in the Quantum Universe* (Chapel Hill: University of North Carolina Press).

43   *Ibid.*, p. 62.

44   For helpful studies of how textual gaps involve the reader, see Wolfgang Iser (1981), *The Act of Reading: A Theory of Aesthetic Response* (Baltimore: Johns Hopkins University Press) and his *The Implied Reader: Patterns of Communication in Prose Fiction from Bunyan to Beckett* (Baltimore: Johns Hopkins University Press, 1974), as well as Goodrich (1986) and Rivero (1991).

I would add that the *systematic* use of self-conscious language, characteristic of meta-fiction, was another thematic and structural aspect of texts that was impacted upon by quantum findings.[45] Waugh's insightful tome falls short of proposing this:

> In a sense, metafiction rests on a version of the Heisenbergian uncertainty principle: an awareness that 'for the smallest building blocks of matter, the very process of observation causes a major disturbance'... and that it is impossible to describe an objective world because the observer always changes the observed. However, the concerns of metafiction are even more complex than this. For while Heisenberg believed one could at least describe, if not a *picture* of nature, then a picture of one's *relation* to nature, meta-fiction shows the uncertainty even of this process.[46]

Bohr is especially cognisant of the limitations of language in physics, and the role of physics as language, as was mentioned earlier. For Bohr, physics is not a portrayal of the world, but 'our way of talking about the world'.[47] Heisenberg attributes the following comment to Bohr: 'when it comes to atoms, language can be used only as in poetry. The poet, too, is not nearly so concerned with describing facts as with creating images and establishing mental connections.'[48] This reinforces how aware of linguistic constraints Bohr was and the similarity he finds between the dilemma faced by quantum physicists and writers in transmitting and representing concepts.

In the annals of the history of ideas, one would have to give credit not just to quantum physicists for linguistic self-reflexiveness but also to Alfred Tarski's philosophical semantics, which predated the linguist Louis Hjelmslev's notion of metalanguage.[49] They helped to develop the concept of using language to refer to language that would become a hallmark device of modern meta-fiction.

The mass media and improved global communications popularised the scientific advances from the first decades of the twentieth century on, as

---

45 The literary precursors of modern metafiction undoubtedly predate the twentieth century, for example, Cervantes's *Don Quijote* or Sterne's *Tristram Shandy*.

46 Patricia Waugh (1984) *Metafiction: The Theory and Practice of Self-Conscious Fiction* (London: Methuen), p. 3.

47 Gregory (1990), p. 95.

48 Bohr (1991a), p. 41.

49 Waugh (1984), p. 4, takes Hjelmslev into account when tracing the development of metalanguage, but not Tarski. Waugh attributes the term 'metafiction' to William Gass (p. 2).

new art was rapidly exchanged between countries, so that intellectual, literary and artistic innovations were widely disseminated, as Friedman and Donley declare.[50] In addition, 'Creative people in all disciplines moved from one centre of activity to another, working in an atmosphere of enthusiastic collaboration, and making the revolution truly international and interdisciplinary'.[51] Philosophy, art, literature and science became reconciled.[52]

More recently, in 1999 the Grupo de Estética Cuántica produced a manifesto called 'Em@ilfesto', which was available on the Internet, 'although the ideas associated with this group began to circulate throughout Spain in the mid-1990's. A variety of discussions of this position have taken place in Spain, in the form of conferences, newspaper articles and public debates', according to Murphy and Caro.[53] Their movement was not limited to Spain: it 'extends worldwide with members in countries such as Argentina, Belgium, [the] USA and Germany, [but] most of the artists and writers live in Spain'.[54]

As Catalá laments, in Latin America,

> *Muchos en épocas anteriores de este siglo, y hoy día, en el mundo hispánico y luso-brasileño, aún viven en un mundo escindido [de las 'dos culturas']. El establecimiento educacional ... se encarga de transmitir este obsoleto modelo. La cultura popular, con los libros de científicos y algunos escritores de habla inglesa, se ha encargado de comenzar a curar esta escisión cultural.[55]*

Such authors as Capra, Talbot, Penrose, Wilber, Bohm and Pagles, for instance, who have popularised the new physics, have been translated into Spanish and are not just read in schools and universities, but also by a broader public in Spanish America.[56]

Selected Spanish American Quantum Fiction

Among others whom I could include, such authors as Borges and Cortázar from Argentina, and Volpi and Fuentes from Mexico, exemplify the rele-

---

50   Friedman and Donley (1990), p. 2.
51   *Ibid.*, p. 2.
52   *Ibid.*, pp. 4, 6.
53   John W. Murphy and Manuel J. Caro, 'Estética Cuántica: A New Approach to Culture,' unpublished ms., p. 1.
54   'Estética Cuántica', p. 14, n. 1.
55   Rafaél Catalá (1998) 'Literatura y ciencia en las culturas de habla espanola,' *La Torre: Revista de la Universidad de Puerto Rico* [Special volume on Hispanic literature and science, Alicia Rivero [Rivero-Potter](ed.)], vol. 3, no. 9, p. 545.
56   *Ibid.*

vance of quantum physics to modern Spanish American narratives. They reflect and transform new ways of understanding reality that the quantum-mechanical findings of Heisenberg provided with respect to uncertainty. The four writers question the epistemological and ontological status of reality as does Heisenberg. They often structure their narratives in ways that violate causality, a characteristic of deterministic, classical physics and realistic narrative. They incorporate the reader actively in their texts and in the creation of textual reality. Readers construct reality in their narratives similarly to physicists: the latter affect reality by using measuring devices that interfere with subatomic particles as the scientists observe them during experiments, which they then interpret. Readers, in turn, alter textual reality by observing, piecing together and interpreting fiction when they (re)read a text.

Quantum physics does not just appear in narrative, but can also be found in other genres, such as poetry, for instance, in the Venezuelan Lucila Velásquez's *La rosa cuántica*,[57] as is obvious from the title, and in the Nicaraguan Ernesto Cardenal's *Cántico cósmico*,[58] which is laden with references to science. In the *Cántico*, Cardenal 'forges Quantum Mechanics into Quantum Poetics', per Lamadrid.[59] Cardenal himself suggests this in 'El cálculo infinitesimal de las manzanas': 'Todo lo que escribo es fragmentario, un conjunto de cuantos'.[60] There are references to quantum mechanics in such poems of the *Cántico cósmico* as 'El big bang', 'En una galaxia cualquiera', 'La danza de los astros' and especially in 'Cántico cuántico'.

Cántico cuántico contains the most examples of quantum mechanics. In it, Cardenal mentions Heisenberg and Bohr, as well as other well-known physicists. The elementary particles are the protagonists of the universal dance of life for Cardenal. He takes into account: the 'incertidumbre cuántica';[61] how the observer brings quantum reality into being; the wave-particle duality and the inability to know simultaneously the position and movement of electrons, among other key, quantum concepts.

With respect to the essay, the Mexican Vasconcelos's 'Ciencia y poesía' in *Ética* joins philosophy, science and literature in the exploration of reality: 'La filosofía supone el aprovechamiento de nuestro más amplio saber para construir la quimera que mejor represente la realidad unificada. Hay en nuestra comprensión de las cosas un mundo poético y un mundo cien-

57   Lucila Velásquez (1992), *La rosa cuántica* (Caracas, Venezuela: Monte Avila).

58   Ernesto Cardenal (1993) *Cántico cósmico* (Madrid: Trotta).

59   Enrique Lamadrid, (1991)'The Quantum Poetics of Ernesto Cardenal, Review of *Cántico cósmico*,' *Ometeca*, vol. 2, no. 2, pp. 147.

60   Cardenal (1993), p. 53.

61   *Ibid.*, p. 239.

tífico: conciliar ambos y hacerlos concurrir a una representación superior y sintáctica: he ahí otra manera de considerar la tarea filosófica.'[62] 'Ciencia y poesía' refers to Jeans and Eddinton — scientists turned philosophers who wrote about the new physics.[63] Quantum physics, relativity and humanities are not only themes in Vasconcelos's *Ética* but also in *La revulsión de la energía*, *Estética* and *Tratado de metafísica*: in them, Catalá remarks that Vasconcelos 'escribe sobre la necesidad de fundir en una sola expresión la física cuántica y la teoría de la relatividad con las humanidades y la religión … Vasconcelos no percibe el universo mecánicamente a la manera newtoniana, sino a la manera de la física cuántica, como idea.'[64]

Borges knew of the new physics, which appears in his texts. He mentions Heisenberg in a footnote to 'M. Davidson: *The Free Will Controversy*' in *Discusión*.[65] In this brief essay, Borges comments on William James — a philosopher who wrote about quantum physics — referring to the notion of free will, which Davidson ignored, according to Borges.[66] Borges proposes modestly that

> *Los deterministas niegan que haya en el cosmos un solo hecho posible, id est, un hecho que pudo acontecer o no acontecer; James conjetura que el universo tiene un plan general, pero que las minucias de la ejecución de ese plan quedan a cargo de los actores… El principio de Heisenberg — hablo con temor y con ignorancia — no parece hostil a esa conjetura.*[67]

Although Bernal is opposed to the implications of quantum mechanics, he succinctly explains how free will can be ascribed to atomic particles: 'it was claimed [by scientific writers and philosophers] that the electron was in a certain sense a free agent. It might or might not at any time do this or that. And if the electron is a free agent why should man not be?'[68]

---

62   José Vasconcelos (1959), *Obras completas*, vol. III (Mexico, DF: Libreros Mexicanos Unidos, 1959), p. 693.

63   José Vasconcelos (1959) 'Ética,' *Obras completas*, p. 695.

64   Catalá (1998) p. 539. Regarding the contribution to intellectual history by philosophical writers like Vasconcelos and others, see Jorge J.E. Gracía and Mireya Camurati (eds) (1989) *Philosophy and Literature in Latin America: A Critical Assessment* (Albany: State University of New York Press).

65   Jorge Luis Borges (1989) *Obras completas: 1923–49*, vol. I (Barcelona: Emecé). I am grateful to Evelyn Fisburn and Eduardo Ortiz for this reference, which she gave him and he sent me in an e-mail, dated 1 August 2000.

66   *Ibid.*, p. 283, n. 1.

67   *Ibid.*

68   J.D. Bernal (1971) *Science in History*, vol. III, (Cambridge, MA: Massachusetts Institute of Technology Press), p. 751.

Gregory summarises the notion of freewheeling particles in quantum physics as follows: 'Since quantum mechanics deals only with probabilities, anything that is not impossible has some probability of occurring. Physicists describe the subatomic world as a freewheeling one where everything that is not prohibited happens at one time or another'.[69] Borges correctly ascribes such a freewheeling antideterminism to Heisenberg and relates it to the notion of free will.

Returning to the narrative genre, Borges uses quantum physical notions in his short stories, for instance, in *El jardín de senderos que se bifurcan*. The structure of the tale's indeterminate, bifurcating outcomes is akin to the idea of quantum free will and not just to the concept of multiple worlds theorised by Wheeler and DeWitt that some critics have cleverly related to this story.[70] The multiple worlds hypothesis itself actually depends on the probabilities of electrons behaving in various ways, albeit it does not put emphasis on free will or freewheeling, but on the observer's interaction with the experimental arrangement, leading to one result or another, which we saw earlier with the double-slit experiment.

As is the case with electrons, many different outcomes are potentially possible, both for the protagonists of Borges's story and for the hero of the paradoxical novel of T'sui Pên contained within it, depending upon what the characters *choose* to do, as James and Heisenberg implied to Borges in *M. Davidson*. Yu Tsun, the protagonist of *El jardín de senderos que se bifurcan*, randomly picked Stephen Albert out of a phone book only because Albert's name fortuitously coincided with the secret location of the Allies' artillery that Yu needed to communicate in code to the Nazis for

69   Gregory (1990), p. 216, n. 49.
70   With respect to multiple worlds in this and other works by Borges, see Michael Capobianco (1989), 'Quantum Theory, Spacetime, and Borges' Bifurcations,' *Ometeca*, vol. 1, no. 1, pp. 27–38; Paul Halpern (1991) 'Borges, Neitzche and Poincaré Recurrence,' *Ometeca*, vol. 2, no. 2, pp. 71–7; Floyd Merrell (1991), *Unthinking Thinking: Jorge Luis Borges, Mathematics and the New Physics* (West Lafayette: Purdue University Press), pp. 177–82; Walter Mignolo (1977), 'Emergencia, espacio, "mundos posibles": Las propuestas epistemológicas de Jorge Luis Borges,' *Revista Iberoamericana*, nos. 100–1, pp. 370–1; Mark Mosher (1994), 'Atemporal Labyrinths in Time: J.L. Borges and the New Physics,' *Symposium*, vol. 48, pp. 51–62. They omit quantum physics, as does Thomas Weissert (1991), 'Representations and Bifurcation: Borges's Garden of Chaos Dynamics,' in Hayles (ed.), *Chaos and Order: Complex Dynamics in Literature and Science* (Chicago: University of Chicago Press), pp. 228, 231, who likens Borges's approach to chaos theory and relativity. Friedman and Donley (1990), pp. 85, 108, include Borges in their discussion of relativity as well as of the quantum joining of subject and object in his *Pierre Menard*, p. 140, but do not treat quantum physics in *El jardín*.

whom he spied. Nonetheless, hypothetically he could also have picked other ways of sending the information. His victim proclaimed that there were many possibilities for the two of them as well, contrary to Newton's temporal determinism, and ironically foreshadows his death, for Yu is not the friend he pretends to be:

> A diferencia de Newton... su antepasado [T'sui Pên] no creía en un tiempo uniforme, absoluto. Creía en infinitas series de tiempos, en una red creciente y vertiginosa de tiempos divergentes, convergentes y paralelos ... No existimos en la mayoría de esos tiempos; en algunos existe usted y yo no; en otros, yo, no usted; en otros, los dos. En éste, que un favorable azar me depara, usted ha llegado a mi casa; en otro, al atravesar el jardín, me ha encontrado muerto; en otro, yo digo estas mismas palabras, pero soy un error, un fantasma ... El tiempo se bifurca perpetuamente hacia innumerables futuros. En uno de ellos soy su enemigo.[71]

Upon hearing Albert's pronouncement, Yu intuits that the garden 'estaba saturado hasta lo infinito de invisibles personas. Esas personas eran Albert y yo ... multiformes en otras dimensiones de tiempo'.[72] Yu ultimately exercises his free will by remorsefully murdering the innocent sinologist, who holds the metafictional keys to unravelling T'sui's labyrinthine novel and Borges's tale. As the observer of this Borgesian system that contains Yu and Albert, Yu goes from being a spectator to an actor. He could have opted to save or to kill Albert; either event was possible until Yu made one of those virtual, quantum probabilities (similar to the 'potentialities' or 'possibilities' mentioned by Heisenberg) real by assassinating Albert.[73]

The importance of the quantum notion of indeterminacy for literature, alluded to as the 'principio de indeterminación' in chapter 4 of Cortázar's Rayuela, is assumed by the narrator without elucidating what it is.[74] That chapter instructs the active reader to skip to chapter 71 in which Heisenberg and Plank are mentioned by Morelli.[75] This self-conscious, narrator-writer of the novel alludes there to atomic randomness with respect to atoms jumping from one energy level to another when excited by heat, light, electricity or radiation, again without illuminating us as to its

---

71   Borges (1989), *Ficciones, Obras completas*, vol. I, p. 479.
72   Borges, *El jardín de senderos que se bifurcan*, p. 479.
73   For a comprehensive examination of Bohr's complementarity and other aspects of Heisenberg's uncertainty and potentialities in *El jardín*, see Rivero [Rivero-Potter] (1997) 'Complementariedad e incertidumbre en "El jardín de senderos que se bifurcan" de Borges,' *La Torre*, vol. 2, no. 6 , pp. 459–74.
74   Julio Cortázar (1974) *Rayuela* (Buenos Aires: Sudamericana), p. 41.
75   *Ibid.*, p. 432.

meaning: Morelli derides journalists who 'se mata[n] explicándonos ... que todo vibra y tiembla y está como un gato a la espera de dar el enorme salto de hidrógeno o de cobalto que nos va a dejar a todos patas para arriba',[76] which implies atomic destruction by a bomb.[77]

Perhaps the same jumping of electrons suggested to Cortázar the hopscotch method with which he challenges readers to actively reconstruct his novel in the 'Tablero de dirección' with which *Rayuela* begins. A clue is provided by his essay, *Algunos aspectos del cuento*, where he says that a good short story revolves around an exceptional theme.[78] He likens said theme metaphorically to a 'sistema atómico', a 'núcleo en torno al cual giran los electrones'.[79] He adds another, even more significant, metaphor: in order for the culmination of the creative process to occur, the meaning of the short story must leap from the originating author to the receiving reader: 'el cuento ... tiene que dar el salto que proyecte la significación inicial, descubierta por el autor, a ese extremo más pasivo ... que llamamos el lector'.[80] In contrast, in chapter 79 of *Rayuela*, the reader is not called upon to be passive (a stereotypical, female 'lector-hembra'), but a participant and an accomplice (a phallocentric, male 'lector... cómplice'),[81] which caused controversy for the author and has been polemical for his critics over the years in terms of how they consider these readers.

Traditional science is often criticised in *Rayuela* by the characters, for it is part of the rational, conventional order they reject in their search for a surreality that science can't encompass. Morelli intends instead to unite

---

76 *Ibid.*
77 Strehle (1992), as well as Friedman and Donley (1990), mention other works by Cortázar with respect to relativity. Sara Castro-Klarán (1983) 'Fabulación ontológica: hacia una teoría de la literatura de Cortázar,' in Jaime Alazraki et al. (eds.), *La isla final* (Madrid: Utramar), pp. 351–72, examines contemporary philosophy and science in Cortázar's works perceptively without including quantum mechanics. Friedman and Donley (1990) relate the quantum joining of subject and object to one of Cortázar's short stories, *Continuidad de los parques*, but they do not equate his techniques to quantum physical concepts in his most famous work, *Rayuela*. Mark Mosher (1996) 'Los "desespacios y los destiempos" en los "wormholes" de Cortázar,' *Hispanófila*, vol. 39, no. 116, pp. 69–82, and Margarita Krakusin (1997) 'Ciencia y arte: el quantum y los agujeros cortacianos en "La isla a mediodía",' *Hispanic Journal*, vol. 18, no. 2, pp. 317–26, discuss the new physics in works by Cortázar from the perspective of wormholes, not of quantum mechanics.
78 Julio Cortázar (1994) *Obra crítica*, vol. II (Madrid: Alfaguara), p. 374.
79 Cortázar, *Algunos aspectos del cuento*, p. 375.
80 *Ibid.*, p. 377.
81 Cortázar (1974), pp. 452–4.

human vision with that of the animal and vegetable kingdom by the 'des-antropomorfización que proponen urgentemente los biólogos y los físi-cos', which he likens to Far-Eastern mysticism.[82] There have been scien-tists such as Capra and Zukav who have established there are similarities between the new physics and Oriental metaphysics.[83] In 'Julios en acción', an essay, Cortázar states that 'en el siglo XX nada puede curarnos mejor del antropocentrismo autor de todos nuestros males que asomarse a la físi-ca de lo infinitamente grande (o pequeño)'.[84]

Generally, in Volpi's *En busca de Klingsor*,[85] quantum physics, and Heisenberg in particular, form an integral part of the novel. There are par-allel plots surrounding Bacon's life and Links's own, which Links relates, in addition to intersections among said plots, as when the two men team up to uncover who was Hitler's right arm, the mysterious German scientist whose code name is *Klingsor*. It is hinted that the nefarious Klingsor might be Heisenberg, given his role in the Nazi atomic program during World War II; this becomes a major theme around which much of the plot re-volves. A plethora of other characters who are fictionalised, real physicists are also mentioned, such as those who worked with Einstein at Princeton University, as well as pro and anti-Nazi German scientists.[86]

Some chapter titles reflect quantum physical notions, such as 'Hipótesis: De la física cuántica al espionaje' and 'El principio de incer-tidumbre'. Planck, Schrödinger, Heisenberg and Bohr appear mentioned in others, as does the act of observing.

The observing reader and narrator are important in Volpi's text and this is especially true in Fuentes's. The reader is more like a detective in Volpi's novel, following clues to attempt to discover who Klingsor is in this open-ended work that does not divulge his identity definitively. Links not only declares that he himself is 'lo que veo', in contrast to the Cartesian *cogito*, but also that truth is relative, misinterpreting Einstein.[87] In addition, his observational role as narrator allows him to complete quantum wave

---

82 *Ibid.*, p. 620.

83 Fritjof Capra (1991)*The Tao of Physics* (Boston: Shambala); Gary Zukav (1979) *The Dancing Wu-Li Masters: An Overview of the New Physics* (New York: William Morrow).

84 *La vuelta al d'a en ochenta mundos* (Mexico, DF: Siglo XXI, 1967), p. 17.

85 Jorge Volpi (1999) *En busca de Klingsor* (Barcelona: Seix Barral).

86 There are technical mistakes in some of the novel's scientific and mathematical de-tails that have been correctly disclosed by Carlos Tello Díaz (1999) *El misterio de Klingsor*, review of *En busca de Klingsor* by Jorge Volpi, *Nexos* [Mexico], vol. 22, no. 264, pp. 90–1.

87 Volpi (1999), p. 24.

states, he claims, and his complement, the reader, should also be an observer, who is to actively fill in textual gaps:

> *Los 'estados de onda' cuánticos que yo completo con mi acto de observación, gracias a[l] … principio de incertidumbre, la teoría de complementaridad, el principio de exclusión — por lo cual nadie puede decir que tiene una verdad mejor que otra.*
>
> *… Recuerden a Schrödinger: para que haya un verdadero acto de conocimiento, debe haber una interacción entre el observador y lo observado, y ahora yo me encuentro en esta segunda … categoría. Disfruten … analizando los efectos que se les presentan y tratando de rastear sus causas. Esta es la clave del éxito científico.*[88]

En busca is a good example of what Strehle calls 'actualist' quantum fiction in *Fiction in the Quantum Universe*, since there is a visibly inventing, metafictional writer-narrator, Links, who is a persona of the author, as were Albert and Morelli, and even more so Cristóbal in Fuentes's novel.

Cristóbal Nonato

Fuentes's unborn narrator, after whom the novel is named, explicitly refers to Heisenberg and Bohr in the novel, as well as to Einstein, Rutherford, Planck, Pauli, Broglie, etc. More importantly, the work is structured on the uncertainty principle. The neonate narrator's scientific grandparents had a photo of Heisenberg, who impressed Cristóbal's father tremendously: 'nada afectó tanto la joven imaginación de mi padre como la certeza de su incertidumbre: la lógica del símbolo no expresa al experimento; *es* el experimento. El lenguaje es el fenómeno y la observación del fenómeno cambia la naturaleza del mismo.'[89] Cristóbal emphasises the linguistic interpretation of the experimental arrangement, but also alludes to the problems of verbally conveying what happens during experiments involving subatomic phenomena. As a narrator-writer and fictional character, Cristóbal fractures conventional syntax and experiments with language in the novel. Words are the stuff of life for him, literally and figuratively.

Cristóbal, who reflects meta-fictionally on his use of words and his task as narrator, declares that observing phenomena changes and creates them. He is a fan of Heisenberg and from his mother's uterus 'observo ya al mundo que me gesta y al cual yo gesto, observádolo … manteniendo abier-

---

88   *Ibid.*, pp. 25, 26.

89   Carlos Fuentes (1987) *Cristóbal nonato* (Mexico, DF: Fondo de Cultura Económica), p. 72.

tas las posibilidades infinitas de observar los infinitos acaeceres del mundo inacabado y transformarlos al observarlos: cambiarlos en historia, narración, lenguaje, experiencia, lectura sin fin'.[90] Cristóbal is the actor/spectator of his uterine drama, as well as an observer of the external world, in addition to observing the fictional world he narrates and invents. While observing it, he selects events from numerous possible ones he could recount, transforming them into language and narrative. Reality is a language game for Fuentes, determined by the narrator and reader, both of whom are observers and fabricators of reality.

Cristóbal's parents, Ángel and Ángeles, at times take turns narrating parts of this polyphonic novel as do some of the other characters (for example, Huevo, who is another mask of the author, as is Cristóbal), or even the reader.[91] Ángel and Ángeles, 'en la noche se pusieron a imaginar probabilidades, alternativas del relato, sin recordar ... que yo mismo poseo mil alternativas'.[92] This implies quantum probability; Cristóbal refers to the plethora of possibilities for narration and narrated events, but also to the genetic chances that he would be born a boy, as his parents desire. Cristóbal adds that there is only instability in a mutable, uncertain universe: 'no hay nada estable, perenne ... todo mutable, mutante'.[93] He views his own body as a system that will act upon the world: 'mi cuerpo/ es el sistema/ con el que voy a contestarle/ al mundo físico, le contestaré al mundo/ creando al mundo, seré el autor de lo que me precede',[94] violating cause and effect.

Events are often shown to have myriad interpretations, which depend not only upon the observing narrator, but also and especially upon the observing reader. Like the observer of a quantum experiment, the reader is not passive, but is an integral part of the system and affects the results. A case in point is the seeming reconciliation of Ángel and Ángeles to his malevolent uncle, Homero, who invites them to celebrate New Year's with him, 'a menos que' and fourteen potential scenarios are listed without any indication of which is the most plausible.[95] There are other examples of multiple versions of events, such as the debatable death of Tomasito, war waged on Mexican soil by the USA with Mexico's collusion, who blame it

---

90   *Ibid.*, p. 72.
91   See Rivero [Rivero-Potter] (1996) 'Columbus's Legacy in *Cristóbal Nonato* by Carlos Fuentes,' *Revista Canadiense de Estudios Hispánicos*, vol. 20, pp. 305–25.
92   Fuentes (1987), p. 251.
93   *Ibid.*, p. 411.
94   *Ibid.*, p. 384.
95   *Ibid.*,, pp. 171–2.

on insurgents, and so on. There are even blank pages that readers are called upon to fill in as they wish, for instance, with respect to Matamoro's text or to Cristóbal's family history.

Cristóbal often alludes or refers explicitly to the act of narrating and reading. On the one hand, a narrator who observes the events of the novel will tinge them subjectively in recounting them — he invents the world he observes, interpreting what he tells. On the other, readers are observers and creators of the novel too, as we have seen. Cristóbal even asserts that 'nada es simultáneo ... todos somos observadores ... es nuestro privilegio, el tuyo y el mío, Elector'.[96] The reader is called an 'Elector' because s/he elects or chooses; the ideal Elector's ludic reading ('Ludectura') entails a most active role — s/he is to be a co-writer ('Autor-Lector') with Cristóbal.[97] During the act of reading, the 'Electores' become its co-authors, for they must assemble their own text from the many versions of events that are presented by a number of narrative voices which blend with that of Christopher.

An interesting strain of this dystopic novel, which adverts to utopia in Pacífica,[98] is that Heisenberg's uncertainty rules in the Eden that is Pacífica, where scientific and spiritual knowledge are joined, as Vasconcelos also envisioned. There, contradictions between time and space, subject and object have been peacefully and fraternally resolved:

La observación de todos los fenómenos simultáneamente es imposible: debemos escoger un tiempo y un espacio dentro del vasto continuo que nos es dado imaginar porque existe en realidad: nuestra rebanada del fenómeno global es nuestro límite pero es nuestra libertad: es lo que podemos afectar, para bien o para mal: lo que podemos ver, tocar, es sólo una cara de la realidad: la posición o el movimiento de algo, uno u otro, pero nunca los dos juntos; éste es nuestro límite pero también nuestro poder: dependemos de la visión del otro para completar nuestra propia visión ... el otro soy yo porque me completa.[99]

---

96   *Ibid.*, p. 560.

97   *Ibid.*, pp. 150–1. On the role of the active, co-authoring reader in Fuentes, see Rivero (1996 and 1991), in addition to Kristine Ibsen (1993), *Author, Text, and Reader in the Novels of Carlos Fuentes* (New York: Peter Lang). William L. Siemens (1991) 'Chaos and Order in Roberto Burgos' El patio de los vientos perdidos and Carlos Fuentes' Cristóbal Nonato,' *Antípodas*, vol. 3, pp. 205–16 relates Heisenberg's uncertainty to reader response, but focuses on chaos theory rather than on quantum mechanics.

98   See Rivero (1996).

99   Fuentes (1987), p. 546.

In the paradise that is Pacífica, the paradoxes inherent in Heisenberg's uncertainty are solved by the completion of one's partial knowledge by another's. Fuentes argues for brotherhood, for the respect of the Other who completes us, like the Platonic half, but does so here in the context of uncertainty. Fuentes thus gives Heisenberg's principle connotations it does not have in physics, of course, but which Fuentes extrapolates from Heisenberg imaginatively.

Again, Fuentes takes liberties with the uncertainty principle, Heisenberg's observer and relativity in the closing chapter of the novel. Cristóbal asks the reader, whom he addresses directly and informally as 'Tú' to remember the photo of Heisenberg in his father's house, 'diciéndonos de despedida que el observador introduce la inseguridad en el sistema porque no puede separarse de un punto de vista por lo tanto el observador y su punto de vista son parte del sistema por lo tanto no hay sistemas ideales porque hay tantos puntos de vista como hay observadores y cada uno ve algo diferente: la verdad es parcial porque la conciencia es parcial: no hay más universalidad que la relatividad'.[100]

As in Volpi's novel, which Fuentes's precedes, truth is partial and relative, something Einstein never proclaimed. Be that as it may, there is a greater multiplicity of observational viewpoints in *Cristóbal Nonato* than in *En busca*, which brings about an even greater uncertainty in the system formed by the text, narrator and reader. Readers participate even more actively in Fuentes's text than in Volpi's.

In conclusion, I have demonstrated that the new physics, especially quantum mechanics, is a valuable tool that can be used to explore understudied dimensions of Spanish American texts. No longer should they be considered in isolation from science or from Anglo-European works with which they have much in common, given the new physics' contribution to the episteme of the twentieth century that revolutionised our modern view of reality and of the arts internationally.

100 *Ibid.*, p. 561.

9

Constructing Postcoloniality:
Scientific Enquiries in *Cien años de soledad*

Patricia Murray

ien años de soledad is a novel about science. Much of the energy, humour and creative enquiry that takes place in the novel is due to the scientific capacities of the various characters and the histories of science that García Márquez is concerned to explore. *Cien años* is also a novel about postcoloniality and as such asks questions about what counts as science and about the boundaries we draw in defining what is, and what is not, a scientific perspective. Mediated largely through the cross-cultural figure of Melquíades, the novel is particularly concerned with notions of where and how science gets produced, how it gets remembered, used and manipulated. There are many lessons along the way. As readers, we are repeatedly encouraged to pluralise our definitions of science, to acknowledge the multiple ways of seeing that are offered in the novel. The characters themselves must often look to marginalised, subaltern sources for guidance in their own scientific enquiries. Most importantly, as I will argue in this chapter, the narrative of scientific progress that takes place in the novel is sharply critical of a modernity that instigates an amnesiac fate. Indeed, loss of memory is a disease that attacks Macondo twice — once as the insomnia plague and again with the banana plague — and the reader is necessarily implicated in the healing process, in that act of memory which becomes central to García Márquez's scientific concerns. In this sense, postcoloniality is always an anticipatory discourse, a future that has still to be written, though it will be my contention that a version of postcoloniality emerges out of the various scientific enquiries that take place in the novel.

My focus here, then, is twofold and I will be concerned to link the themes of science and postcoloniality throughout. After an initial exploration of the role of science in the novel, paying particular attention to the play and complexity of textual form, I will proceed with a brief note on the aspects of postcoloniality that are of interest to me, also highlighting the role of Melquíades in linking the two areas of concern. I will then move through the novel chronologically, focusing on five successive case studies to reveal aspects of scientific enquiry pertinent to my argument: (1) José Arcadio Buendía's use of the magnet to extract gold; (2) the plague of insomnia; (3) modern inventions as ambiguous tools of progress; (4) examination of local bananas and the subsequent banana plague; (5) exegesis of the manuscripts.

I

Science is everywhere in García Márquez's *Cien años de soledad.* The novel is immersed in, is structured by and is constantly commenting on our changing notions of science and scientific perspectives. From the very first sentence, when Aureliano remembers that distant afternoon 'en que su padre lo llevó a conocer el hielo' ['when his father took him to discover ice'],[1] García Márquez reminds us of the personal discovery that science is. The reference to ice is surprising, incongruous, even comic for the reader who is well used to its everyday existence, but this does not detract from the sense of wonder conveyed as the Buendías first make contact with the invention. The fact that Aureliano's father 'interrumpió la lección de física'(p. 73) ['had interrupted the lesson in physics' p. 20] in order to take him to the spectacle, underlines the scientific rediscovery that we are making - that the block of ice is also a lesson in physics. In this sense *Cien años* immediately questions our notion of what counts as science, and the boundaries we draw in defining our own scientific knowledge.

How do you respond, for instance, to José Arcadio's stargazing and cosmological enquiries and his eventual revelation that 'La tierra es redonda como una naranja'(p. 62) ['The earth is round like an orange', p. 12]? Again, García Márquez's style ensures the initial comedy, playing to the contrast between the craziness of new scientific insights (Ursula's response) and the banality of their routine familiarity (reader response). But in the level of descriptive detail García Márquez is also commenting on the extent of scientific endeavour (concrete evidence of the earth's shape first came from watching the stars) and reminds us of the great efforts of imagination that were needed to grapple with and understand the implications of a spherical earth. There is admiration, a recreation of the 'wonder' of science that can be mind-boggling and overwhelming, also a lesson in different scientific traditions. José Arcadio makes this discovery after it has

---

1   Gabriel García Márquez, *Cien años de soledad* [1967](Madrid: Espasa-Calpe, S.A., 1985), p. 59; *One Hundred Years of Solitude*, translated by Gregory Rabassa [1970] (London: Picador, 1978), p. 9; all future quotations are from these editions.

been proven elsewhere (Melquíades arrives to inform them of this), just as Europe came to such knowledge after it was known in the Arabic world.

Despite José Arcadio's initial fear that 'Aquí nos hemos de pudrir en vida sin recibir los beneficios de la ciencia'(p. 70) ['We're going to rot our lives away here without receiving the benefits of science', p. 18], Macondo acts as a meeting place for scientists of various traditions and times, allowing García Márquez to comment on the collaborative nature of scientific exchange and to contrast some very different cultural paradigms. In order to pluralise our notions of science, the first few chapters are packed with different kinds of scientific endeavour and activity. The first travellers to Macondo, for instance, the family of gypsies led by Melquíades, pitch up their tents to display new inventions. Science here is performance, accompanied by music and theatre, and the villagers are entertained and startled by the spectacle of moving iron created by the magnets and the changes in perspective brought on by the telescope and the magnifying glass. José Arcadio Buendía's attempts to subject each new invention to a utilitarian end,[2] like turning the magnifying glass into a weapon of war, are the source of much humour for the reader and an early example of García Márquez satirising a certain western scientific paradigm:

> *... entregado por entero a sus experimentos tácticos con la abnegación de un científico y aun a riesgo de su propia vida. Tratando de demostrar los efectos de la lupa en la tropia enemiga, se expuso él mismo a la concentración de los rayos solares y sufrió quemaduras que se convirtieron en úlceras y tardaron mucho tiempo en sanar ...*

> *Pasaba largas horas en su cuarto, haciendo cálculos sobre las posibilidades estratégicas de su arma novedosa, hasta que logró componer un manual de una asombrosa claridad didáctica y un poder de convicción irresistible* (p. 61).

[... completely absorbed in his tactical experiments with the abnegation of a scientist and even at the risk of his own life. In an attempt to show the effects of the glass on enemy troops, he exposed himself to the concentration of the sun's rays and suffered burns which turned into sores that took a long time to heal... He would spend hours on

---

2   See Floyd Merrell's (1989) article 'José Arcadio Buendía's Scientific Paradigms: Man in Search of Himself,' in Harold Bloom (ed.), *Modern Critical Views: Gabriel García Márquez* (New York: Chelsea House Publishers), pp .21–32, which contrasts the utilitarian and non-utilitarian modes of thinking of JAB and Melquíades respectively and shows how the transmutations in JA's conception of nature are analogous to the development of scientific thought in the western world.

end in his room, calculating the strategic possibilities of his novel weapon until he succeeded in putting together a manual of startling instructional clarity and an irresistible power of conviction, p. 10.]

For José Arcadio, method (theorem-proof, theorem-proof) is everything and his excessive reliance on this is contrasted with the sensory capacities of some of the other characters, like the premonitions experienced by Ursula and Aureliano or Pilar Ternera's ability to read the future in the cards. Nevertheless, it is José Arcadio's belief in the progress of science and his devotion to scientific enquiry which inspires the friendship with Melquíades, in many ways the key relationship in the novel. As Melquíades moves into the Buendía household we see a proliferation of scientific spaces, from the noisy laboratory in which the two friends collaborate, to the expert silversmith workshop of Aureliano and the frenetic activity of the masons and carpenters as Ursula meticulously supervises the redesigning of the house. Subsequent chapters introduce us to further examples of energetic, and often eccentric, scientific innovators, such as Pietro Crespi, Aureliano Triste and Aureliano Centeno and Gaston. From the experimental speculations of the abstract thinker to the hand of the skilled artisan, then, the point is well made that science and scientific perspectives are always everywhere.

This is not, of course, a world in which the numinous, the phenomenological is absent. The frequent recurrence of unexplained phenomena — the rain of tiny yellow flowers on José Arcadio's death, the numerous ghosts and spirits present in the novel, Amaranta's meeting with death — as well as all the marvellous half-truths and exaggerations, are central to the spirit and texture of the novel. Even towards the end we are confronted by startling images that challenge our own scientific perspectives and reading strategies — like the steel twine that grows around Nigromanta's waist and the mysterious angel that appears in Macondo. Only Melquíades 'parecía conocer el otro lado de las cosas' (p. 63) ['seemed to know what there was on the other side of things', p. 12], in fact we are told that his tribe 'había sido borrada de la faz de la tierra por haber sobrepasado los límites del conocimiento humano' (p. 94) ['had been wiped off the face of the earth because they had gone beyond the limits of human knowledge', p. 39]. For the rest of us, a great deal is still unknown and García Márquez's excursions into the marvellous real are a comment on the kinds of methodology he proposes for open-minded, intuitive interpretation. The strict rational criteria of José Arcadio become redundant in such circumstances, like his attempt to use the daguerreotype 'para obtener la prueba científica de la existencia de Dios' (p. 108) ['to obtain scientific proof of the existence of God', p. 50]:

Mediante un complicado proceso de exposiciones superpuestas tomadas en distintos lu-gares de la casa, estaba seguro de hacer tarde o temprano el daguerrotipo de Dios, si existía, o poner término de una vez por todas a la suposición de su existencia (p. 108).

[Through a complicated process of superimposed exposures taken in different parts of the house, he was sure that sooner or later he would get a daguerreotype of God, if He existed, or put an end once and for all to the supposition of His existence, p. 51.]

A spiritual dimension is validated in the novel, but it cannot be measured in this way. Nor are priests the arbiters of the sacred in García Márquez's world. The humorous dialogue between José Arcadio and Father Nicanor Reyna is an example of the usefulness of scientific enquiry for García Marquez, especially in counteracting the claims of religion. Father Nicanor initially tries to instil some faith in José Arcadio, but he rejects the 'vericuetos retóricos' ['rhetorical tricks'] of the priest and the 'objetos artesanales sin fundamento científico' ['artistic objects without any scientific basis']:

Pero entonces fue José Arcadio Buendía quien tomó la iniciativa y trató de que-brantar la fe del cura con martingalas racionalistas ... Cada vez más asombrado de la lucidez de José Arcadio Buendía, le preguntó cómo era posible que lo tuvieran amarrado de un árbol.

- Hoc est simplicisimun — contestó él —porque estoy loco.

Desde entonces, preocupado por su propia fe, el cura no volvió a visitarlo ... (p. 137).

[But then it was José Arcadio Buendía who took the lead and tried to break down the priest's faith with rationalist tricks ... Even more startled at José Arcadio Buendía's lucidity, he asked him how it was possible that they had him tied to a tree.

'Hoc est simplicissimus,' he replied. 'Because I'm crazy.'

From then on, concerned about his own faith, the priest did not come back to visit him... p. 75.]

José Arcadio's scientific paradigms lack the intuitive and spiritual insights of Melquíades and Ursula, and this is a balance which is important to García Márquez. But his ability to ask questions and his desire for information, and change, make him a willing adversary of all forms of authority, including the Church. In the end it is his own rage for order and common-sense rationality that defeats him. From his careful observation of nature he arrives at a second moment of epiphanic insight:

> La máquina del tiempo se ha descompuesto' — casi sollozó ... Pasó seis horas examinando las cosas, tratando de encontrar una diferencia con el aspecto que tuvieron el día anterior, pendiente de descubrir en ellas algún cambio que revelara el transcurso del tiempo (p. 132).

> ['The time machine has broken,' he almost sobbed ... He spent six hours examining things, trying to find a difference from their appearance on the previous day in the hope of discovering in them some change that would reveal the passage of time, p. 71.]

But his realisation of the relativity of time and space brings on a delirium and he is tied to a tree. Although later generations of Buendías acknowledge his lucidity and scientific insight, José Arcadio is allied with a mechanistic view of the universe, with Newtonian methods and perspectives, and he is unable to move with the New Physics into a post-Einsteinian world view.

II

The novel invites us, also, to read through changing scientific perspectives and in relation to insights gained from the New Physics. The complexity, and play, of the novel's structure and form can usefully be related to its status as quantum fiction;[3] fiction which is, like the quantum universe, relative, uncertain, complementary. The interconnection of space and time, for instance, posited by Einstein in his special theory of relativity (1905) as a space-time continuum, introduces us to the notion of a four-dimensional reality in which time is the fourth dimension. As Gary Zukav writes, this is a difficult world to visualise:

> If we could view our reality in a four-dimensional way, we would see that everything that now seems to unfold before us with the passing of time, already exists in toto, painted, as it were, on the fabric of

---

3   See Susan Strehle's (1992) book, *Fiction in the Quantum Universe* (University of North Carolina Press) and her summary of Einstein's special (1905) and general (1916) theories of relativity, Heisenberg's uncertainty principle (1927) and Niels Bohr's principle of complementarity (1927), pp. 9–14. Although Strehle confines her analysis to six North American texts her methodology can usefully be applied to a reading of *Cien años*.

space-time. We would see all, the past, the present and the future with one glance.[4]

This is exactly the challenge posed by Melquíades' manuscripts and one of the reasons why it takes so many generations of Buendías to translate them into the book we are reading:

Melquíades no había ordenado los hechos en el tiempo convencional de los hombres, sino que concentró un siglo de episodios cotidianos, de modo que todos coexistieran en un instante (p. 446).

[Melquíades had not put events in the order of man's conventional time but had concentrated a century of daily episodes, in such a way that they coexisted in one instant, p. 335.]

If we could unravel such a labyrinth completely then we would perceive the quantum moment as described by Zukav. As it is, we have a partial version of Melquíades' text, multiply translated into the chaotic novel of *Cien años de soledad*. We can trace the influence of Borges here and his own scientific/metafictional obsessions, specifically in 'El jardín de senderos que se bifurcan' ['The Garden of Forking Paths']. In this story we are told of Ts'ui Pen:

... docto en astronomía, en astrología y en la interpretación infatigable de los libros canónicos, ajedrecista, famoso poeta y calígrafo: todo lo abandonó para componer un libro y un laberinto.

[... a man learned in astronomy, astrology, and the unwearying interpretation of canonical books, a chess player, a renowned poet and calligrapher — he abandoned it all in order to compose a book and a labyrinth.][5]

He retreats to the 'Pabellón de la Límpida Soledad'(p.109) ['Pavilion of Limpid Solitude' p. 81] to complete his task and now, 'al cabo de más de cien años'(p. 110) ['more than a hundred years after the fact', p. 82], a man named Stephen Albert has realised that the book and the labyrinth are one

4   Gary Zukav (1990 [1979]) *The Dancing Wu Li Masters: An Overview of the New Physics* (London: Rider).

5   Jorge Luis Borges (1987 [1944]) 'El jardín de senderos que se bifurcan' (1941), collected in *Ficciones* (Madrid: Alianza Editorial), p.109; translated by Andrew Hurley as 'The Garden of Forking Paths' in *Fictions* (Penguin, 2000), p. 81.

and the same — *'el jardín de senderos que se bifurcan* era la novela
caótica'(p.111) ['the garden of forking paths was the chaotic novel' p.83]
— and begins to translate this contradictory and confusing novel that is
predicated upon 'infinitas series de tiempos, en una red creciente y vertig-
inosa de tiempos divergentes, convergentes y paralelos'(p. 114) ['an infinite
series of times, a growing, dizzying web of divergent, convergent and parallel times', p. 85].

Melquíades is a worthy inheritor of Ts'ui Pen's art and García
Márquez's novel contains many examples of the kinds of narrative fork-
ings suggested in Borges' story. At the beginning of chapter 6, for instance,
we are told of Colonel Aureliano Buendía:

> *Tuvo diecisiete hijos varones de diecisiete mujeres distintas, que fueron exterminados*
> *uno tras otro en una sola noche, antes de que el mayor cumpliera treinta y cinco años*
> *(p. 155)*

[He had seventeen male children by seventeen different women and
they were exterminated one after the other on a single night before the
oldest one had reached the age of thirty-five, p. 91.]

Only to be told in chapter 12 that on the night of the extermination, one
of his sons, Aureliano Amador:

> *logró saltar la cerca del patio y se perdió en los laberintos de la sierra que conocía*
> *palmo a palmo gracias a la amistad de los indios con quienes comerciaba en*
> *maderas. No había vuelto a saberse de él (p. 283).*

[had been able to leap over the wall of the courtyard and was lost in the
labyrinth of the mountains, which he knew like the back of his hand
thanks to the friendship he maintained with the Indians, from whom
he bought wood. Nothing more was heard of him, p. 198.]

This last piece of information is then again contradicted in chapter 18
when the aging Aureliano Amador returns to Macondo after his long years
as a fugitive, goes unrecognised by the later generation of Buendías, and is
finally shot by the secret agents who have chased him.

Like the Ts'ui Pen of Borges' story, García Márquez creates several fu-
tures, several times, which themselves proliferate and fork. What happens,
for instance, to Meme and Mauricio Babilonia after they are separated and
how do you read their love story? At the end of chapter 14 we are told with
assurance that Mauricio 'murió de viejo en la soledad, sin un quejido, sin

una protesta, sin una sola tentativa de infidencia' (p. 330) ['died of old age
in solitude, without a moan, without a protest, without a single moment of
betrayal', p. 238]. Uncertainty is introduced in the next chapter when we
are told that Meme 'admitió como una verdad irremediable, que Mauricio
Babilonia había muerto' (p. 333) ['admitted as an irremediable truth that
Mauricio Babilonia had died', p. 241). When she enters the convent it is her
own loyalty in love that is emphasised:

> Todavía pensaba en Mauricio Babilonia, en su olor de aceite y su ámbito de mari-
> posas, y seguiría pensando en él todos los días de su vida, hasta la remota madru-
> gada de otoño en que muriera de vejez ...(p. 334).

> [She was still thinking about Mauricio Babilonia, his smell of grease,
> and his halo of butterflies, and she would keep on thinking about him
> for all the days of her life until the remote autumn morning when she
> died of old age... p. 242.]

This version is then contradicted by Aureliano Babilonia's interpretation of
events at the end of the novel, when he is finally able to translate
Melquíades' prophetic text:

> ...y encontró el instante de su propia concepción entre los alacranes y las mariposas
> amarillas de un baño crepuscular, donde un menestral saciaba su lujuria con una
> mujer que se le entregaba por rebeldía (p. 447).

> [... and he found the instant of his own conception among the scor-
> pions and the yellow butterflies in a sunset bathroom where a mechanic
> satisfied his lust on a woman who was giving herself out of rebellion,
> p. 335.]

Reading is always also an interpretation and Aureliano Babilonia's must be
contextualised in terms of the trauma he has just suffered. The deliberate
use of unreliable narration, the relative and uncertain twists typical of
quantum fiction, must also alert us to the dangers of reading the ending
too literally, despite its assured tone. In the same way as complementarity
sees irreconcilable and mutually exclusive concepts — light as both particle and wave — as necessary to understand subatomic reality (which be-
haves according to both opposite principles), so we must bring the same
flexibility to our interpretation of the quantum text. In one scenario
Melquíades' manuscripts are turned into sawdust, in another they are the
book we are reading, and each re-reading reverses the conclusion that
'todo lo escrito en ellos era irrepetible desde siempre y para siempre' (p.

448). ['everything written on them was unrepeatable since time immemorial and forever more', p. 336].

In the end there is too much humour and energy for the novel to be read pessimistically. After all, as Aureliano Babilonia discovers, 'la literatura fuera el mejor juguete que se había inventado para burlarse de la gente' (p. 421). ['literature was the best plaything that had ever been invented to make fun of people', p. 314.] and there are many jokes along the way. I am reminded of the 'butterfly effect' — the swarm of yellow butterflies that follow Mauricio Babilonia causing a storm elsewhere — a reference to the work of Edward Lorenz (1961) and almost a pun on Chaos Theory. But if García Márquez is keen to draw attention to the fabrication of it all (as in the collapse of the brothel into a half-sketch), to break the spell of the narrative, it is in order to force the reader out of the fictionality of Macondo into a renewed awareness of their own responsibility in shaping a future that will learn from these lessons of the past.

III

As a novel about postcoloniality, *Cien años* looks forward to the end of isolation and the possibility of a future that contains all of the diversity encapsulated in Macondo as well as avoiding some of the reasons for its decline. On the one hand, Macondo was a prosperous place until 'lo desordenó y lo corrompió y lo exprimió la compañía bananera' (p. 383) ['it was disordered and corrupted and suppressed by the banana company', p. 282] and there is a critique of neo-colonialism throughout the novel. But the narrative is also concerned with questions of migration and hybridity and with the failure of the Buendía family in relation to these concerns. In this sense, *Cien años* is also about the colonial hierarchies that the Buendías and others have internalised and which continue to frustrate their attempts at community.

The second chapter tells us that in the sixteenth century, an Aragonese merchant comes to live in a settlement of peaceful Amerindians and does business there with a native-born ('criollo') tobacco planter. These are the ancestors of Úrsula Iguarán and José Arcadio Buendía, who are thus born, as it were, out of Ortiz's counterpoint of tobacco and sugar.[6] Rather than embracing the process of transculturation that this implies, however, successive generations of Buendía and Iguarán become ever more closed and

6   See Fernando Ortiz (1940) *Contrapunteo cubano del tabaco y el azúcar* (Havana) in which he metaphorises Cuban identity as existing in the counterpoint between tobacco (signifying native, black, sacred, etc) and sugar (signifying foreign, white, capitalistic, etc).

interbred, the family line becoming infinitely predictable and haunted by the fear that pervades the novel — of the birth of a child with a pig's tail. On the one hand, it is significant that many of the Buendía men[7] have important relationships with women from subaltern groups — José Arcadio (Jr) with the gypsy from the circus, Aureliano Segundo with the mulatta, Petra Cotes, and Aureliano Babilonia with the Afro-Caribbean, Nigromanta. It is also significant, however, that they do not marry any of these women. Even García Márquez's favourite character of Petra Cotes — 'el único nativo que tenía corazón de árabe' (p. 367) ['the only native who had an Arab heart', p. 269] — must remain the mistress while Fernanda del Carpio, 'una cachaca' (p. 359) ['a stuck-up highlander', p. 263] plays the part of the wife. This failure on the part of the Buendías to embrace a hybrid future, preferring instead to turn inwards (even incestuously) and backwards (to the repressive codes of a Spanish colonial past in the figure of Fernanda), is as responsible for Macondo's decline as the corruption and brutality of the banana company.

In the context of nineteenth-century scientific racism which provides the backdrop to many of the Buendías' lives, and its obsessive classifications of the 'degenerative' results of racial hybridisation,[8] it is possible to discern in the narrative the outline of a vicious pigmentocracy that continues to favour light over darker skin colour. It is significant, for instance, that the narrative voice (which is often saturated in the perspectives and prejudices of the characters at the time) comments on Rebeca being lighter, and prettier, than Amaranta and contrasts the fair-skinned Remedios Moscote (the preferred bride) with her sister Amparo, who is 'morena como su madre' (p. 112) ['dark like her mother', p. 54].

Contrasted with such neurosis is the character of Macondo itself. Situated close to the Caribbean coast of Colombia and surrounded by various Amerindian groups, Macondo is constantly defined in opposition to the gloomy, icy, stuck-up, highland regions (home of the capital and an emerging national elite) which García Márquez portrays as overly-controlled by the corrupt figures of Church and State. Initially founded by the Buendías and other migrating families, Macondo becomes a trope for mi-

---

7   The women, of course, are forbidden to do the same. Only Meme transgresses class and racial hierarchies in her passion for Mauricio Babilonia (the mechanic from the town who, significantly, looks like he is a descendant of the gypsies) and this brief relationship is ended by Fernanda's murderous intervention.

8   See discussion by Robert Young (1995) in *Colonial Desire: Hybridity in Theory, Culture and Race* (Routledge), pp. 175–82.

gration itself welcoming first the gypsies, then successive generations of Arabs, Afro-Caribbeans, some Guajiro and other Amerindians, political refugees like the wise Catalonian as well as composite nomadic figures like Melquíades and the Wandering Jew. This richly diasporic and multi-lingual contact zone resists all attempts by the centre to impose a homogenous national identity, until the arrival of the military junta which enforces the law of the banana company in a final, brutal wave of neo-colonial exploitation.

Though marginalised by economic power structures, Macondo's subaltern peoples are often central to its rehabilitation and García Márquez is keen to acknowledge their status and claims to knowledge, as well as their role in reclaiming historical and scientific memory. It is to these sources that later generations of Buendías must turn as they pursue the translation of the manuscripts and the quest to understand their own identity in the world. Significantly, our chief decoders in this regard are themselves examples of criminalised and marginalised voices — José Arcadio Segundo, champion of workers' rights and a victim of state-sponsored massacre, and Aureliano Babilonia, the illegitimate, hybrid, locked-away son of Meme and Mauricio Babilonia. Aureliano is frequently referred to as 'un antropófago' ['a cannibal'], and like the hero of Andrade's manifesto[9] he devours all knowledge, becoming as proficient in Papiamento as in Sanskrit, to become our main conduit to memory.

IV

It is Melquíades, of course, who guides both renegade Buendías in their final scientific-philosophical task, as indeed he guides José Arcadio Buendía in his initial scientific explorations. Omnipresent in the novel, Melquíades is truly able to exist in four-dimensional space-time; constantly rebirthing, it seems, from our limited three-dimensional perspectives. His mother tongue is Sanskrit and he journeys from North West India through the Middle East and North Africa before arriving in Spain and crossing to Latin America. These are journeys in time as well as space so that we can only guess at the translations of his name and the hybridising of identity that have taken place before we meet this composite Arabic/gypsy figure who operates as the key subaltern influence and source of learning in the novel. On the one hand, he is a fantasy of post-coloniality: a nomadic wanderer who writes in his tribe's original sacred lan-

9   See Oswald de Andrade (1928), 'Manifesto Antropófago', translated by Leslie Barry as 'Cannibalist Manifesto,' in *Latin American Literary Review*, vol. 19, no. 38, pp. 38–47.

guage (Sanskrit) though he speaks 'un intrincado batiburrillo de idiomas' (p. 125) ['a complex hodgepodge of languages', p. 65]; a scientist-seer who retains multiple histories of scientific and philosophical exchange, despite being a refugee who 'sufría por los más insignificantes percances económicos' (p. 63) ['suffered from the most insignificant economic difficulties', p. 13]. On the other, he is a reminder of the exclusions of modernity, of the eclipsed perspectives that postcolonialism must learn to remember.

V

But if *Cien años de soledad* (a title that speaks of the past) is a novel about postcoloniality (a term that implies the future), how do we bridge these distinct temporal zones to create, in García Márquez's words, a present day in which 'the lineal generations of one hundred years of solitude will have at last and for ever a second chance on earth?'[10] In other words, what sorts of postcoloniality, or strategies for achieving postcoloniality, does García Márquez propose? At this point I want to bring together the two themes of my chapter to show how a version of postcoloniality emerges out of the various scientific enquiries that take place in the novel.

The novel begins with a warning, and it is one that crucially links the narratives of colonialism and capitalism. The introduction of the magnet, contextualised in a playful and animistic way by Melquíades and the other gypsies, inspires a dream of riches in José Arcadio who 'pensó que era posible servirse de aquella invención inútil para desentrañar el oro de la tierra' (p. 60) ['thought that it would be possible to make use of that useless invention to extract gold from the bowels of the earth', p. 9]. Although Melquíades warns him that it will not work for this purpose José Arcadio presses on, trading his only assets (a mule and a pair of goats from which his wife earns a small income) for the magnetised ingots and justifying the risk in terms of future excess — 'Muy pronto ha de sobrarnos oro para empedrar la casa' (p. 60) ['Very soon we'll have gold enough and more to pave the floors of the house', p. 9]. Although José Arcadio's attempts to 'demostrar el acierto de sus conjeturas' (p. 60) ['demonstrate the truth of his idea', p. 9] are conveyed in typically comic style, the link with a previous, excessive pursuit of instant wealth and folly is made explicit:

10   Gabriel García Márquez (1987) 'The Solitude of Latin America: Nobel Address 1982' [translated by Richard Cardwell], in McGuirk and Cardwell (eds) *Gabriel García Márquez* (Cambridge University Press), p. 211.

Lo único que logró desenterrar fue una armadura del siglo XV con todas sus partes soldadas por un cascote de óxido, cuyo interior tenía la resonancia hueca de un enorme calabazo lleno de piedras. Cuando José Arcadio Buendía y los cuatro hombres de su expedición lograron desarticular la armadura, encontraron dentro un esqueleto calcificado que llevaba colgado en el cuello un relicario de cobre con un rizo de mujer. (p. 60)

[The only thing he succeeded in doing was to unearth a suit of fifteenth-century armour which had all of its pieces soldered together with rust and inside of which there was the hollow resonance of an enormous stone-filled gourd. When José Arcadio Buendía and the four men of his expedition managed to take the armour apart, they found inside a calcified skeleton with a copper locket containing a woman's hair around its neck, pp. 9–10.]

The reference to the 'enorme calabazo' renders an image of nature untouched within man's hubris and although the presence of the locket around the now absent identity of the desiring conquistador primarily points to the folly of his expedition, it implies, typically of the author, that the romantic impulse also survives, though in richly ambivalent circumstances.

This initial portrait of José Arcadio as scientific explorer thus provides a stark reminder of colonialism's desiring machine and the 'delirium for gold'[11] which has ravaged so much of the continent's past. Amerindian peoples, of course, bear the brunt of this encounter and it is significant that when Cataure and Visitación, a Guajiro prince and princess, arrive in Macondo they are 'huyendo de una peste de insomnio que flagelaba a su tribu desde hacía varios años' (p. 93) ['in flight from a plague of insomnia that had been scourging their tribe for several years,' p. 38] Visitación, now a servant in the Buendía household, explains to José Arcadio how the sickness leads to a debilitating loss of memory:

Quería decir que cuando el enfermo se acostumbraba a su estado de vigilia, empezaban a borrarse de su memoria los recuerdos de la infancia, luego el nombre y la noción de las cosas, y por último la identidad de las personas y aun la conciencia del propio ser, hasta hundirse en una especie de idiotez sin pasado (p. 99).

[She meant that when the sick person became used to his state of vigil, the recollection of his childhood began to be erased from his memory, then the name and notion of things, and finally the identity of people

---

11   See García Márquez's Nobel Address, pp. 207–8.

and even the awareness of his own being, until he sank into a kind of idiocy that had no past, p. 43]

This erasure of cultural memory, dramatised here as a physical illness but linked metaphorically to the assaults of colonialism, is a theme that runs throughout the novel and the act of memory, the ability to 'see' clearly (into the future as well as the past) becomes central to García Márquez's scientific concerns. On this occasion the Amerindians' privileged knowledge is not heeded: 'José Arcadio Buendía, muerto de risa, consideró que se trataba de una de tantas dolencias inventadas por la superstición de los indígenas' (p. 99). ['José Arcadio Buendía, dying with laughter, thought that it was just a question of one of the many illnesses invented by the Indians' superstitions,' p. 43.] With the predictably colonialist response of the 'moderniser', José Arcadio dismisses the 'backwardness' of the Amerindians and disregards their medical learning (and historical experience) as ineligible scientific perspectives.

When the dreaded plague inevitably begins to infect Macondo, García Márquez charts a variety of scientific responses to it. Ursula, who is associated with popular and oral traditions that also lie on the margins of enlightenment respectability, tries to utilise her knowledge of the medicinal value of plants. José Arcadio and the other elders devise a system of quarantine so that the infection will not spread to other towns. Aureliano 'concibió la fórmula que había de defenderlos durante varios meses de las evasiones de la memoria' (p. 102) ['conceived the formula that was to protect them against loss of memory for several months', p. 45] — the system of labelling objects, animals, concepts, methods which they are quickly forgetting. The increasingly western scientific language that García Márquez uses, however — 'El letrero que colgó en la cerviz de la vaca era una muestra ejemplar...' (p. 102) ['The sign that he hung on the neck of the cow was an exemplary proof...' p. 46] — only serves to underline the one-eyed, inappropriate reliance on such methods, culminating in the Borgesian absurdity of José Arcadio's memory machine:

El artefacto se fundaba en la posibilidad de repasar todas las mañanas, y desde el principio hasta el fin, la totalidad de los conocimientos adquiridos en la vida. Lo imaginaba como un diccionario giratorio que un individuo situado en el eje pudiera operar mediante una manivela, de modo que en pocas horas pasaran frente a sus ojos las nociones más necesarias para vivir (p. 103).

[The artefact was based on the possibility of reviewing every morning, from beginning to end, the totality of knowledge acquired during one's life. He conceived of it as a spinning dictionary that a person placed on the axis could operate by means of a lever, so that in very few hours there would pass before his eyes the notions most necessary for life, p. 46.]

As the limitations of José Arcadio's scientific paradigms are again exposed, the ancient figure of Melquíades returns and is immediately able to locate the medicine that will cure the town of their loss of memory. There is a double movement taking place here. Melquíades signifies a specifically Eastern knowledge and as such displays an integrated awareness of the traditional sciences of medicine, law, theology and philosophy. The act of memory that is central to García Márquez's scientific concerns, thus, also means remembering the multiple histories of science, and specifically the role of the East in the development of western scientific methods. As Floyd Merrell has pointed out,[12] the scientific inventions (the astrolabe, the compass, the telescope and the magnifying glass) that Melquíades introduces to Macondo are all Arabic contributions to the western world from the tenth to the twelfth centuries and José Arcadio's intellectual transformation as a result of this contact mirrors the enormous, though not always acknowledged, influence of the Arabic world on European versions of enlightenment.

But in his role as healer, Melquíades also has many of the attributes of the shaman, the traditional priest-doctor of the Amerindians. The shaman combined both science and ritual in his role as spiritual healer and it is in this dual capacity that Melquíades befriends and guides José Arcadio. The shaman reserved his strongest medicine for the most hazardous path of all, that through death, which he would have travelled himself at least once. In similar fashion, Melquíades returns from the dead many times to guide successive generations of Buendías in their attempts to recover the past and learn something of their own identity. But before we look at this most important of scientific enquiries — exegesis of the manuscripts — we must make a fuller journey through the repeating cycles of the novel.

The tenth chapter of *Cien años* triggers a strange feeling of déjà vu in the reader. The repetition of grammatical constructions and observations from the opening chapter produces the chaotic sense of time having somehow stretched and folded, so that we seem to be retracing many of the same journeys, though with slight differences. There is so much initiative, imagination and scientific enterprise again that Ursula begins to echo the feelings of the reader: 'Ya esto me lo sé de memoria ... Es como si el tiempo diera vueltas en redondo y hubiéramos vuelto al principio' (p. 240).

12   Merrell (1989), pp. 22–3.

['I know all of this by heart… It's as if time had turned around and we were back at the beginning', p. 162.] Although no-one remembers the 'empresas colosales' (p. 140) ['colossal undertakings', p. 162] of José Arcadio anymore, the character of Aureliano Triste is described as having the same kind of modernizing zeal as his grandfather. He too wants to link the town with the rest of the world and it is his construction of the railroad which enables the arrival of that most ambivalent symbol of scientific progress:

El inocente tren amarillo que tantas incertidumbres y evidencias, y tantos halagos y desventuras, y tantos cambios, calamidades y nostalgias había de llevar a Macondo (p. 266).

[The innocent yellow train that was to bring so many ambiguities and certainties, so many pleasant and unpleasant moments, so many changes, calamities and feelings of nostalgia to Macondo, p. 184.]

The train brings 'tan maravillosas invenciones' (p. 267) ['such marvellous inventions', p. 185] that the inhabitants of Macondo are immediately reminded of the gypsies. These modern inventions, however, are described as ambiguous tools of progress — the electricity is too noisy, for instance, and the phonograph is a mere mechanical trick that cannot compare with a band of musicians:

Fue una desilusión tan grave, que cuando los gramófonos se popularizaron hasta el punto de que hubo uno en cada casa, todavía no se les tuvo como objetos para entretenimiento de adultos, sino como una cosa buena para que la destriparan los niños (p. 268).

[It was such a serious disappointment that when phonographs became so popular that there was one in every house they were not considered objects for amusement for adults but as something good for children to take apart, p. 185.]

The creolising of the scientific invention as described here, and the dismissal of those 'equilibristas del comercio ambulante' (p. 268) ['ambulatory acrobats of commerce', p. 186] who fail to impress with such consumerist necessities as a whistling kettle, are symptomatic of Macondo's sense of its own history and its ability to bring a questioning, scientific perspective to the foreigners' inflated claims of modernity.

One particular scientific enquiry changes all this, however. Mr Herbert's meticulous examination of the local bananas at the Buendía table:

Entonces sacó de la caja de herramientas que siempre llevaba consigo un pequeño estuche de aparatos ópticos. Con la incrédula atención de un comprador de diamantes examinó meticulosamente un banano seccionando sus partes con un estilete especial, pesándolas en un granatorio de farmacéutico y calculando su envergadura con un calibrador de armero. Luego sacó de la caja una serie de instrumentos con los cuales midió la temperatura, el grado de humedad de la atmósfera y la intensidad de la luz (p. 269).

[Then he took a small case with optical instruments out of the toolbox that he always carried with him. With the suspicious attention of a diamond merchant he examined the banana meticulously, dissecting it with a special scalpel, weighing the pieces on a pharmacist's scale, and calculating its breadth with a gunsmith's callipers. Then he took a series of instruments out of the chest with which he measured the temperature, the level of humidity in the atmosphere and the intensity of the light. p. 187.]

This is followed by the unnatural excavations of the land by engineers, agronomists, hydrologists, topographers and surveyors and leads to the unleashing of a plague every bit as debilitating as that of insomnia — 'la peste del banano' (p. 273) ['the banana plague', p. 190]. Recalling the delirium for gold alluded to in José Arcadio's first scientific exploration, the bananisation of Macondo becomes a violent, exploitative, capitalist enterprise referred to by García Márquez simply as that 'eructo volcánico' (p. 273) ['great volcanic belch', p. 190].

Science here is a central dynamic in the colonial machine that repeats itself in the form of the plantation.[13] Without even masquerading as a symbol of liberty, progress and universal reason (its usual contradictory performance as an instrument of empire), the science machine enables the banana company to change the pattern of the rains, accelerate the cycle of harvests and move the river to the other side of town, in its desire to maximise profits from the banana plantations. In its repressive despotism, the science machine (to follow Benítez-Rojo's model) is also a naval machine, a military machine, a bureaucratic machine, a commercial machine, a political machine, a legal machine. The workers, for instance, are not paid in real

13   According to Benítez-Rojo's reading, the plantation is the 'strange attractor' of the circum-Caribbean, the central pull which produces an endlessly proliferating and interconnected series of colonial machines. See his discussion of this in 'From Columbus's Machine to the Sugar-Making Machine,' in *The Repeating Island: The Caribbean and the Postmodern Perspective*, 2nd edition (Duke University Press, 1996 [Spanish language edition 1989]), pp. 5–10.

money but in scrip which can only be exchanged for Virginia ham in the company commissaries. José Arcadio Segundo is thrown into jail because he reveals that:

> ... *el sistema de los vales era un recurso de la compañía para financiar sus barcos fruteros, que de no haber sido por la mercancía de los comisariatos hubieran tenido que regresar vacíos desde Nueva Orleáns hasta los puertos de embarque del banano* (p. 338).

> [... the scrip system was a way for the company to finance its fruit ships, which without the commissary merchandise would have to return empty from New Orleans to the banana ports, pp. 244–5.]

The legal machine quashes the workers' protests by establishing that 'the workers' do not exist, making it easier for the military machine to then massacre those workers who are on strike in the central square. The train, that symbol of ambivalent modernity, carries out the piles of dead bodies — 'quienes los habían puesto en el vagón tuvieron tiempo de arrumarlos en el orden y el sentido en que se transportaban los racimos de banano' (p. 344) ['those who had put them in the car had time to pile them up in the same way in which they transported bananas', p. 250] — even as the political machine is wiping out all memory of the striking workers and the science machine is bringing on the storms and hurricanes that will last for four years, eleven months and two days.

During the frenetic times of the banana company, we are told that the only serene corner had been established by:

> ... *los pacíficos negros antillanos que construyeron una calle marginal, con casas de madera sobre pilotes, en cuyos pórticos se sentaban al atardecer cantando himnos melancólicos en su farragoso papiamento* (p. 271).

> [... peaceful West Indian Negroes, who built a marginal street with wooden houses on piles where they would sit in the doors at dusk singing melancholy hymns in their disordered gabble, p. 188.] [14]

After the deluge, it is noticeably the third generation of Arabs, descendants of those who 'hallaron en Macondo un buen recodo para descansar de su milenaria condición de gente trashumante' (p. 367) ['had found in Macondo a good bend in the road where they could find respite from their

---

14   The West Indians are Papiamento creole speakers, a specificity which is lost in the translation.

age-old lot as wanderers', p. 269] who appear on the streets to enjoy their first sunshine. These markers of postcoloniality, suggesting multiple histories of migration and transculturation, are important antidotes to the one-sided, conquistadorial perspectives that dominate the world of the novel. Although, like the Amerindians, they are marginal to the social structures of Macondo, they are always positive signifiers for García Márquez and indicative of an erasure that is still to be fully written.

It is José Arcadio Segundo, the sole survivor of the massacre, who begins that quest for postcolonial self-identity. Having escaped from the train of death, it is fitting that he finds respite in the supernatural glow of Melquíades' room. The room itself is an authorial comment on the ability to see clearly, on the act of memory that is about to be embarked upon and the perspectives necessary for that journey. When the soldier who searches the room does not see 'los ojos árabes' (p. 348)[15] ['the Arab eyes' p. 253] of José Arcadio Segundo — as he sits beside the fresh ink and dust-free parchments of the old scientist-seer — he repeats the limited, materialist perspective of the later Colonel Aureliano Buendía, who also saw only the debris of a snake-infested room that had been uninhabited for one hundred years. These soldiers cannot see that in this room it is always a Monday in March, fresh and clean as the day Melquíades arrived, because they cannot 'see' the marvellous, their senses now so dulled that they are no longer susceptible to such surprises of the imagination. José Arcadio Segundo, however, 'el habitante más lúcido de la casa' (p. 383) ['the most lucid inhabitant of the house', p. 283], comes to realise that:

> ...*también el tiempo sufría tropiezos y accidentes, y podía por tanto astillarse y dejar en un cuarto una fracción eternizada* (p. 384).

[... time also stumbled and had accidents and could therefore splinter and leave an eternalised fragment in a room, p. 283.]

Armed with such heightened quantum perspectives, and guided by the presence of Melquíades, José Arcadio Segundo devotes himself to the most important scientific enquiry of all, the study of the manuscripts. By

---

15   This particular description at this point prepares us for the link that is about to be forged between José Arcadio Segundo and Melquíades. However, pointing to the Arabic features of some of the Buendías is also García Márquez' way of reminding us that the Spanish are already always hybridised, and in particular are already Africanised (through contact with North African Arabic cultures) before they reach the Americas.

the time of his death, he succeeds in deciphering the logic of an alphabet that had bewildered so many others:

Las letras parecían ropa puesta a secar en un alambre, y se asemejaban más a la escritura musical que a la literaria (p. 230).

[The letters looked like clothes hung out to dry on a line and they looked more like musical notation than writing, p. 154.]

He thus bequeaths to his apprentice great-nephew both the keys to translate the Sanskrit language and the formula of his own traumatic memory:

Acuérdate siempre de que eran más de tres mil y que los echaron al mar (p. 388).

[Always remember that there were more than three thousand and that they were thrown into the sea, p. 287.]

The young Aureliano preserves this political memory, even when people think it is 'una versión alucinada' (p. 383) ['a hallucinated version', p. 283] because it radically disagrees with the false one consecrated by historians. He also avidly studies all of the books that make up Melquíades' library. At first, this seems like another of García Márquez's hyperbolic jokes:

Aureliano no abandonó en mucho tiempo el cuarto de Melquíades. Se aprendió de memoria las leyendas fantásticas del libro desencuadernado, la síntesis de los estudios de Hermann, el tullido; los apuntes sobre la ciencia demonológica, las claves de la piedra filosofal, las centurias de Nostradamus y sus investigaciones sobre la peste, de modo que llegó a la adolescencia sin saber nada de su tiempo, pero con los conocimientos básicos del hombre medieval (p. 390).

[Aureliano did not leave Melquíades' room for a long time. He learned by heart the fantastic legends of the crumbling books, the synthesis of the studies of Hermann the Cripple, the notes on the science of demonology, the keys to the philosopher's stone, the *Centuries* of Nostradamus and his research concerning the plague, so that he reached adolescence without knowing a thing about his own time but with the basic knowledge of a medieval man, p. 288.]

But it soon becomes clear that it is one of the author's serious jokes. In contrast with Fernanda's kind of outdated, mannered, aristocratic learning that proves entirely useless, Melquíades' learning is part of a storehouse of knowledge that Aureliano needs in order to decipher the manuscripts, which are his own and the Buendías' history. When the time is right, Melquíades directs Aureliano to the wise Catalonian's bookshop — a bookshop of 'incunabula'[16] — where he finds the remaining texts he needs to complete his scientific-philosophical task, an understanding of his own identity in the world.

As with many of the scientific enquiries that have taken place in the novel, Aureliano Babilonia must look to subaltern sources for guidance, sources that speak of the intellectual histories of his own ancestors and of the diverse peoples that contribute to the making of Macondo; sources that have been marginalised by the European colonial model of cultural periodisation that fixes notions of 'enlightenment' and 'modernity' in terms of its own perceived centrality. It is not that European learning is absent. The six-volume English Encyclopaedia, for instance, is invaluable in confirming the language of the manuscripts as Sanskrit. But by focusing on patterns of scientific and philosophical exchange that are embodied in the nomadic figure of Melquíades, and then investing in that figure something of the Amerindian shaman, García Márquez succeeds in decentring and provincialising the colonial machine of Europe and validating other resources and perspectives that are also available.

The dangers of isolation and of looking inward are only too evident in the demise of the Buendía family. But the narrative of scientific progress that takes place in the novel is also critical of a modernity that simply eclipses previous traditions and multiple ways of seeing. In order to see clearly, in the truly clairvoyant sense of some of the characters, García Márquez encourages us to look into the future as well as the past, to find in the text:

> ... *la posibilidad científica de ver el futuro transparentado en el tiempo como se ve a contraluz lo escrito en el reverso de un papel* ... (p. 424).

> [... the scientific possibility of seeing the future showing through in time as one sees what is written on the back of a sheet of paper through the light ... pp. 316–7.]

So that although the novel ends at the close of another cycle — the birth of a child with a pig's tail and the cyclonic destruction of Macondo — it also remains poised for the start of the next. García Márquez proposes a version of postcoloniality that is politicised and of long-memory but, above all, one that has still to be written.

---

16   Books printed at an early date, especially before 1501 [from Latin incunabula meaning 'swaddling-clothes, cradle'].

VI

Escaping the criticism García Márquez levelled at his earlier work as 'libros que acaban en la última página'[17] [books which end on the final page], the narrative of *Cien años* forks in several directions after the novel's close. Continuing to read through changing scientific perspectives it is possible, for instance, to trace the contours of a richly multi-layered and fractal angel in the way the sickly creature (who has had its winged chopped off in Macondo) appears after the rains in *Un señor muy viejo con unas alas enormes* (1968) [A Very Old Man with Enormous Wings]. Mandelbrot coined the term *fractal* from the Latin adjective 'fractus' (meaning broken) and 'fractional', and invented fractal geometry as a way of studying highly complex and irregular forms that are common in nature but often misread as counter-intuitive and monstrous.[18] The ambiguous anatomy of the old man causes the priest of Macondo to declare him a monster intent on evil, though his captors are less certain how to measure his form:

> *Tenía el cuerpo cubierto de una pelambre áspera, plagada de garrapatas menudas, y el pellejo petrificado por una costra de rémora, pero al contrario de la descripción del párroco, sus partes humanas eran más de ángel valetudinario que de hombre, porque las manos eran tersas y hábiles, los ojos grandes y crepusculares, y tenía en los omoplatos los muñones cicatrizados y callosos de unas alas potentes, que debieron ser debastadas con hachas de Labrador, p. 379.*

[Its body was covered with rough hair, plagued with small ticks, and the skin was hardened with the scales of a remora fish, but unlike the priest's description, its human parts were more like those of a sickly angel than of a man, for its hands were tense and agile, its eyes large and gloomy, and on its shoulderblades it had the scarred-over and calloused stumps of powerful wings which must have been chopped off by a woodsman's axe, p. 279.]

Summarily disposed of in the highly condensed interlude in which he appears in *Cien años*, it is not until the enquiries of the story that we can ap-

---

17   See Gabriel García Márquez (1982) *El olor de la guayaba: conversaciones con Plinio Apuleyo Mendoza* (Barcelona: Bruguera), p. 82

18   See Katherine Hayles' (1990) discussion 'Strange Attractors: The Appeal of Chaos,' in *Chaos Bound: Orderly Disorder in Contemporary Literature* (Cornell University Press), specifically pp. 163–6.

preciate, with the doctor, 'la lógica de su sus alas'[19] [the logic of his wings]. The supplement of the story goes on to suggest a potential curative healer that is not recognised in Macondo; the intertextual links between the two confirming our own sense of infinite exegesis.

---

19   Gabriel García Márquez, 'Un señor muy viejo con unas alas enormes,' collected in *La increíble y triste historia de la cándida Eréndira y de su abuela desalmada* (Mondadori, 1992), p. 17.

10

Holograms and Simulacra:
Bioy Casares, Subiela, Piglia

Geofferey Kantaris

> Había que tener cuidado al enfrentar un delirio
> de simulación, [...] por ejemplo el de los locos
> furiosos capaces de fingir docilidad o el de los
> idiotas capaces de simular gran inteligencia. [...]
> Nunca se sabe si una persona es inteligente o si
> es un imbécil que *finge* ser inteligente.
> – *La ciudad ausente* (p. 14).

Introduction

I begin with the controversial premise that science as such no longer exists, or at least that it has waned, having been swallowed up and instrumentalised in technology and into the production of the social more generally. Which is to say that we are all, in a sense, post-scientific subjects; that a pure science which is not already bound to technification, the production of informational space and of virtual reality, is no longer possible. What this chapter aims to do is to chart this movement from science as investigation of the real — as empirical discovery — to science as one of the primary discourses by which the real, the reality *effect*, is now produced, simulated, replicated. I shall do this by examining the inscription of science and technology in three Argentine cultural texts spanning some 52 years of the twentieth century: Adolfo Bioy Casares' novella *La invención de Morel* (1940), Eliseo Subiela's film *Hombre mirando al sudeste* (1986) and Ricardo Piglia's novel *La ciudad ausente* (1992).[1]

The conundrum which I hope to present to you, if not to provide answers for, is this: why do all three texts imagine technology very specifically in its relationship to the production of simulacra? And, secondly, why does

---

1   *La invención de Morel*, El Libro de Bolsillo #393 (Buenos Aires: Alianza Emecé, 1991 [1940]); *Hombre mirando al sudeste* (Argentina: Cinequannon, 1986) Colour/35mm/100mins; *La ciudad ausente*, 3rd edition, Biblioteca Breve (Buenos Aires: Seix Barral, 1995 [1992]).

each text seem compelled to think the simulacrum through a grid of political intrigue that entails crime, evasion, discipline and punishment? To
put it succinctly, what is the structural relationship between a regime of
simulation and the individual tactic of dissimulation? But first it is necessary to look briefly at a fundamental precursor to any attempt to think
technology in its relationship to the imaginary, the work of Roberto Arlt
and the role of the Inventor.

The Inventor

In a sense the story of the interrelation between science, technology, simulation and the State in Argentine culture begins inevitably with Arlt. His
work is present throughout the thought of one of my authors, Ricardo
Piglia, and Argentina's prime cultural analyst, Beatriz Sarlo, gives him a
major role in her meticulous study of the technical imagination in
Argentine modernity, *La imaginación técnica: sueños modernos de la cultura argentina*.[2] Arlt provides us with several keys for linking art and simulation, invention and falsification.

Sarlo claims that up until Arlt, modernity had not been thought in its
technological dimension in Argentine culture:

> *[E]ntre los intelectuales, la modernidad no había sido pensada en su dimensión tec-
> nológica. Arlt piensa precisamente esa dimensión. Por eso su imaginación es a la vez
> tan extraña y tan irreconocible si se la compara con la de sus contemporáneos de la
> literatura culta. Abre una utopía tecnológica que, como toda utopía, es un proyecto
> de cambio radical de las condiciones presentes, pero, sobre todo, diagnóstico de aque-
> llo que el discurso postula y construye como realidad.*[3]

That technology should be so intimately bound up with a discourse that
postulates and *constructs* the real, and that literature should explore this technological dimension as a diagnostic, is a first important key to reading the
mode of techno-scientific simulation in Bioy Casares, Subiela and Piglia.

But Sarlo points out another theme which will be of fundamental importance in understanding the relationship of science and technology to
the creative imagination, and this is the fundamental ambiguity of the technological dream. That technology and discourses of science can be the
very underpinning of the authoritarian state and, *at the same time*, the source

2   Beatriz Sarlo (1997 [1992]), *La imaginación técnica: sueños modernos de la cultura argentina*,
     2nd edition (Buenos Aires: Nueva Visión).
3   *Ibid.*, p. 64.

of a counter-praxis and a counter-aesthetic, is a key idea in all of these texts in one form or another:

> *Lo que es instrumento de una sociedad autoritaria (y enloquecida en su autoritarismo) es al mismo tiempo material de ensoñación y fuente de belleza. Arlt escribe un capítulo de la estética industrial, que tiene posibilidades mortíferas pero produce identificaciones y fantasías poéticas. Erdosain sueña un sueño tecnológico donde la destrucción y la belleza se enlazan en una ambigua relación de necesidad. [...] El paisaje industrial, que cubre con metal, rieles y cables todo resto de naturaleza es bello.*[4]

The figure of the small-time inventor is a key trope in all of the texts I am examining, and a fundamental one for thinking this ambiguity of technology and scientific discourse. The inventor is the technological artist, the one who has a chance, however infinitesimal, of shuffling the codes of a consolidating discourse of technocratic power. And as Sarlo's study reveals, the Argentina of the 1920s and 1930s seems to abound with small-time inventors, placing faith in the power of their imagination to change the real through a chance technological discovery.

But it is Ricardo Piglia himself, an indefatigable reader and re-reader of Arlt as well as of Macedonio Fernández, who points out the precise relationship between invention, simulation and the submission of the social to the ferocious fiction of money. Arlt's inventors do not believe in the system, rather they want to play the system for all it is worth, their power of invention aimed at producing the fake and the forgery which will exploit capitalism's ability to transform a fundamental fiction — that of money — into a reality effect. The figure of the copper rose comes to stand as a symbol of this technological artifice:

> *Los personajes de Arlt no tratan de ganar dinero sino de fabricarlo. [... I]nventar es una operación demiúrgica destinada a encontrar la piedra filosofal moderna, el oro que no lo es: la rosa de cobre. Todas las máquinas, los laboratorios, las fórmulas, los aparatos que circulan en la obra de Arlt tienen como objetivo [la] producción imaginaria de riqueza. [...] Inventores, falsificadores, estafadores, estos 'soñadores' son los hombres de la magia capitalista: trabajan para sacar dinero de la imaginación.*[5]

So, hopefully, we now have some keys for reading the links between the image of dissimulation as faking and a system that increasingly produces the real through technological simulation.

---

4   *Ibid.*, p. 57.
5   Ricardo Piglia (1993) *La Argentina en pedazos*, Colección Fierro (Buenos Aires: Ediciones de la Urraca), p. 125.

From Dissimulation to Simulation

La invención de Morel

In 1940, Adolfo Bioy Casares published a short novel, *La invención de Morel*, which charts, in very prescient manner, a journey from terrorised dissimulation to post-technological simulation. A strange rewriting and transformation of H.G. Wells's *The Island of Dr. Moreau* of 1896,[6] it concerns a political refugee, escapee from life imprisonment on a trumped-up charge in Caracas, who makes his way to a secret and uninhabited southwest Pacific island. Although the island is rumoured to be the focus of some strange disease, as yet unknown to science, which kills from the outside in, causing the body to lose its substance and crumble slowly away, he prefers to take his chances there than to live a life of dissimulation and hiding, pursued by the world's police forces for a crime he did not commit.

What he finds there is a terrifying and seductive new technological invention, created by a mad scientist Morel, a 1940s televisual answer to Wells' Moreau who, we may recall, fashioned human beings out of pumas in a double allegory of science and colonisation, medicine and mission. Morel's invention also transforms bodies, quite literally consuming the real to produce its hyperreal simulacrum. In that, it can serve for us today as an allegory for a new kind of technologically mediated colonisation — that of globalisation — and precursor of a new mode of telematic simulation.

The island contains some abandoned buildings: a museum, a chapel and a swimming pool (the relationship of the figure of the museum to the production and containment of reality effects is one that is hinted at here and taken up extensively by Piglia, as we shall see). Believing himself to be alone on the island, the narrator survives by eating plant roots, until one night he awakens to find people — a group of holidaymakers, he thinks — suddenly on the island. A fugitive from justice, he has to hide from them, but is nevertheless intrigued by them, particularly by one woman, who he later finds out is called Faustine. The intrigue of the story lies in the slow and paranoia-inducing discovery by the narrator that none of the people on the island are real: they are all projections of a new recording device, precursor of virtual reality, invented by Morel. Morel himself appears as one character in this simulation of an entire week in the lives of the holidaymakers, a week which, the narrator discovers, is recorded on a self-repeating disk; a week destined to be repeatedly projected in all its details, in every act, movement and thought, for eternity, or for as long as the

6   H.G. Wells (1896) *The Island of Dr. Moreau* (New York: Garden City Publishing, 1896).

machines, powered by tidal hydroelectricity, continue to function. The projections are not merely images: Morel has discovered how to project matter itself into space, but the holograms cannot interact with anything that was not already in the recording, for they are not in that sense alive, repeating mechanically the actions, words and even, it is conjectured, thoughts that they underwent at the time that they were recorded. Morel explains to his gathered guests at the end of the recorded week, an explanation that is itself contained within the recording:

> *Mi abuso consiste en haberlos fotografiado sin autorización. Es claro que no es una fotografía como todas; es mi último invento. Nosotros viviremos en esa fotografía, siempre. Imagínense un escenario en que se representa completamente nuestra vida en estos siete días. Nosotros representamos. Todos nuestros actos han quedado grabados.*[7]

He then explains the functioning of his invention, in a passage that would be cited textually in Eliseo Subiela's film of 1986, as we shall se:

> *Me puse a buscar ondas y vibraciones inalcanzadas, a idear instrumentos para captarlas y transmitirlas. [...] Esta es la primera parte de la máquina; la segunda graba; la tercera proyecta. No necesita pantallas ni papeles; sus proyecciones son bien acogidas por todo el espacio y no importa que sea día o noche. En aras de la claridad osaré comparar las partes de la máquina con: el aparato de televisión [...]; la cámara que toma una película de las imágenes traídas por el aparato de televisión; el proyector cinematográfico.*[8]

But the device, this ultimate holographic projection, has a terrible cost. In recording matter itself, it extracts some fundamental essence of matter, leaving the original that was recorded without material substance. Between eight and 15 days after having been recorded by the machine, organic bodies begin to decay from the outside in. The body, still alive, begins to crumble, consumed, literally, by the image, until the original dies, leaving only the holographic projection to repeat for eternity.

Published in 1940, this vision of simulation consuming the real is a remarkably prescient allegory for an effect that, many would argue, only begins to make itself felt fully from the 1960s, and which Jean Baudrillard analyses some 40 years later as the 'liquidation of all referentials' in the age of simulation. Inverting a parable by Borges at the beginning of his essay on 'Simulacra and Simulations', it is now the real, Baudrillard suggests, whose shreds slowly rot away beneath our simulations, the real 'whose ves-

---

7   Bioy Casares ([1940] 1991), p. 80.
8   *Ibid.*, pp. 84–5.

tiges subsist here and there, in the deserts which are no longer those of Empire, but our own. *The desert of the real itself.*[9]

The protagonist of *La invención de Morel*, literally seduced by the simulacrum of a woman, decides to record himself next to her, simulating interaction with the holidaymakers, although the interaction is merely an illusion, eternally replayed as the simulation of a dissimulation. At the close of the novel, he writes of his body disintegrating, accompanied by what he believes must be atrocious pain, but which is somehow numbed and from which he is totally detached, while his image lives on in the projection.

Hombre mirando al sudeste

In 1986, three years after the dissolution of a ferocious dictatorship in Argentina, Eliseo Subiela makes a film, *Hombre mirando al sudeste*, in which the collapse of a boundary between dissimulation and simulation seems once more to be at stake. A patient calling himself Rantés appears in an asylum in Buenos Aires claiming that he is from outer space. More specifically, he claims to be the projection, or hologram, of some giant extra-terrestrial computer. The job of the doctor, whose name is Julio Denis, is to unmask Rantés' delusion, to put the reality principle firmly back in place, while the film's job would appear to be the steady erosion of the doctor's certainty, and with it our own, about the patient's ontological status. In a *Marat/Sade* type allegory, it is, bizarrely, Julio Denis, alias Julio Cortázar,[10] who finds himself playing Pontius Pilate, indeed ultimately torturer and executioner, to Rantés' cybernetic Christ. The asylum is, amongst other things, an allegory of the Argentine State, dependent for its operation on the submission of minds and bodies to a violent, drug-enforced reality principle. Rantés, 'un simulador' in the doctor's words, seems to reveal the sham nature of the State's own simulation of the real, and pays for it through a replay of the Christian passion.

I shall now analyse an extended sequence from *Hombre mirando al sudeste* which addresses both directly and through various levels of intertextual reference the knot which ties science, technology and the production of simulacra within political regimes of 'truth'.[11] The sequence begins with

---

9   Jean Baudrillard (1988 [1981]) 'Simulacra and Simulations,' in Mark Poster (ed.), *Jean Baudrillard: Selected Writings*. (Cambridge: Polity), pp. 166–7.

10   Julio Denis was the pseudonym under which Cortázar published his earliest literary works. This reference combines with a number of cultural allusions in the film: Magritte, Proust, the Bible, Philip K. Dick and Bioy Casares.

11   The sequence can be viewed at http://www.latin-american.cam.ac.uk /simulacra/clips/.

Dr Denis asking Rantés the name of his mother; this, we may recall, is the key (Oedipal) question which catches out the replicant Léon in Ridley Scott's 1982 film *Blade Runner*, since the biologically engineered replicants in that film do not have mothers, conventional Oedipal histories, or fully-fledged 'human' emotions.[12] Rantés goes on to explain his ontological status as a hologram:

Rantés: *No tenemos madres, o por lo menos, nunca nos hablaron de eso.*

Denis: *¿Es un robot?*

R: *No. Ustedes son robots y todavía no se dieron cuenta. [...] Es que ustedes están en la prehistoria de los hologramas.*

D: *¿Hologramas?*

R: *Sí. Una especie de fotografía obtenida a través de un rayo láser. Es un experimento que suele hacerse en los laboratorios de física. Nosotros hemos logrado ... ¿cómo explicarle? ... que esas imágenes se corporicen en el espacio a través de lo que sería un ... gran proyector, programado con una computadora muy compleja, que incluye en ese rayo, todos los datos vitales para que esa imagen tenga vida. [...] Digo [imagen] para que usted me entienda. Somos réplicas humanas perfectas, salvo por una cosa: no podemos sentir.*

There are at least three overt references to *Blade Runner* in this dialogue, but more importantly (and this is also the intention behind that film), the dialogue and subsequent action serve to throw into doubt the ontological basis for distinguishing between reality and simulation on several levels. By claiming that he is a mere projection or hologram, Rantés is in one very real sense telling the truth. For us as cinema spectators he is, of course, just such a projection, a cinematic image on a screen, yet for us he also is not a mere projection, for it is a fact we must disavow in order to carry on believing in the cinematic illusion, just as it is a fact which Dr Denis must disavow if he is to continue having faith in the psychiatric clinic and the reality principle which it supports. Rantés continually inverts the orders of hallucination and reality: '*Usted* tiene halucinaciones,' he tells the doctor. 'Yo soy una de sus halucinaciones'. In this way, the film cleverly plays on cinematic disavowal in order to reveal the wider disavowals which authorise what counts as reality and reason within the social contract to which the doctor subscribes: or as Rantés puts it at a later point, 'su realidad es espantosa, doctor'.

12 Ridley Scott (1982), *Blade Runner* (USA, Colour/70mm/117mins, based on the 1968 novel by Philip K. Dick, *Do Androids Dream of Electric Sheep*). The exception is the replicant Rachael who has had implanted childhood memories.

Continuing with this sequence, Dr Denis, unable to resist his curiosity as to the source of this particular delirium, goes to check out Rantés' story in a physics laboratory. The scientist provides him with a full demonstration and explanation of the functioning of holograms and assures him that it would be possible to project the perfect illusion of a human being 'en base a láseres pulsados'. In voice-over, the doctor comments, '¿Por qué suponer que alguien que habla de un fenómeno físico tiene que ser un físico?', while on the visual track we see Rantés' fingerprints being taken by the police back in the asylum. In this case, the visual diegesis serves as an ironic counterpoint to the audio track, suggesting that psychiatric science and the state security apparatus are in the business of controlling what can count as reality within the social contract. It is worth noting that, as we shall see in *La ciudad ausente*, science and in particular physics plays an equivocal role by seeming here to confuse, in its findings, the very truth effects for which it acts as discursive guarantor within technocratic political and economic systems. Back at home, Dr Denis discovers 'un eco literario' in Rantés' delirium, reasoning that 'el que manipula [la] información [científica] con fines no científicos [...] es un escritor [...] o simplemente un lector'. Rummaging amongst piles of books on his floor, he picks out *La invención de Morel* by Adolfo Bioy Casares and begins to recite: 'Me puse a buscar ondas y vibraciones inalcanzadas, a idear instrumentos para captarlas y transmitirlas...'

La ciudad ausente

> *Todos vamos a terminar así,*
> *una máquina vigilando a otra máquina.*[13]

The Baudrillardian theme of the proliferation of simulacra also underlies another text, published in 1992, Ricardo Piglia's *La ciudad ausente*, which has been variously described as a 'futurist/cyberpunk detective story',[14] a novel of social and political mourning and the restitution of memory, a homage to Macedonio Fernández and James Joyce, and a postmodern machine for generating stories. It is set in 2004 or 2005 ('hace quince años que cayó el Muro de Berlín'),[15] in and around an occupied Buenos Aires, an emptied-

---

13   Piglia (1992), p. 157.
14   Idelber Avelar (1999), *The Untimely Present* (Durham and London: Duke University Press), p.16.
15   Piglia (1992), p. 144.

out globalised city where time has been standardised worldwide: 'Habían unificado la hora en todo el mundo para coordinar las noticias del telediario de las ocho. Tenían que vivir de noche, mientras en Tokyo salía el sol'.[16]

It is a dense and baffling hallucinatory text in which the re-reading, displacement, re-combination and invention of stories is figured as the only means of resistance to a post-dictatorial, technocratic police state, a.k.a. free-market democracy, which has abolished memory and history to allow for the free circulation of commodities, controls the populace through televisual feedback — a technologically-mediated, quasi-telepathic monitoring of thoughts — and institutes a regime of 'truth' and 'transparency', where the police are the agents of a ferocious reality principle and psychiatrists are the inheritors of the techniques and know-how of erstwhile torturers (in this, Piglia's approach is very similar to that of Subiela):

> *El comisario sonrió. Querían controlar el principio de realidad.*
>
> *[...] —La policía —dijo— está completamente alejada de las fantasías, nosotros somos la realidad y obtenemos todo el tiempo confesiones y revelaciones verdaderas. Sólo estamos atentos a los hechos. Somos servidores de la verdad.*[17]

Piglia shares with Jean Baudrillard a view of the postmodern, free-market, globalised, capitalist polis as engaged in a ferocious, compulsively re-iterated simulation of the real. To the extent that it was the business of the dictatorships to impose the social conditions that would allow the transition from State to Market in the Southern Cone, the democratic governments that marked the end of dictatorship are in some sense the inheritors and continuers of the work done by the military, rather than representing an absolute break with 'authoritarianism', which is how they projected themselves ideologically. In agreement with Idelber Avelar in his book *The Untimely Present* on postdictatorial Latin American fiction, I take this to be one of the central ideas underlying *La ciudad ausente*, or as the cyborg storytelling machine puts it in her final Molly Bloom-esque monologue in the novel:

> *el liberalismo, las tasas libres liquidaron el negocio [de los contrabandistas], el fin del contrabando [...] es el fin de la historia argentina. Ésa era una novela río, empezaba en 1776, en las dos orillas del Plata, la chalupa con las mercancías inglesas y ahora se terminó, tantos muertos para nada, tanto dolor.*[18]

---

16 *Ibid.*, p. 72.
17 *Ibid.*, pp. 94–6.
18 *Ibid.*, p. 166.

The stories, constituting much of the novel, which are figured as circulat-
ing clandestinely throughout Buenos Aires and which are so threatening to
the regime of truth upheld by the police and the psychiatrists in the novel,
are transmitted by this cyborg woman-machine imagined by the avant-
garde Argentine writer 'Macedonio Fernández' as a way of keeping alive
the memory of his beloved Elena: 'Macedonio no intentaba producir una
réplica del hombre, sino una máquina de producir réplicas. Su objetivo era
anular la muerte y construir un mundo virtual'.[19] This machine, enclosed
by the State within a museum, was constructed with the help of one Emil
Russo (the reference to Rousseau's 1762 work on education, *Émile*, is obvious), an Arltian style inventor, possibly a Hungarian collector of automata, or he might perhaps be a Swiss physics teacher called Richter, who
spun the fib to Perón that he was an atomic physicist refugee from Nazi
Germany, selling him the 'secret' of nuclear cold fission woven out of hot
air. Perón, we are told, spent a small fortune on this dream of the
Argentine atom bomb: 'Richter se infiltró en el Estado argentino, infiltró
su propia imaginación paranoica en la imaginación paranoica de Perón'.[20]
Although eventually revealed to be different, Russo and Richter are confused by several characters in the novel, and the confusion points us, I
think, to a knot of political philosophy and scientific discourse underpinning the truth effects and truth claims of the modern information State.

The novel seems to suggest that it is possible to subvert this techno-scientific system, by introducing into it instabilities and paradoxes, all of them
occasioned by the proliferation of recombinatory, self-referential fictions,
told by 'la máquina de Macedonio'. In other words, exploiting the ambiguity
noted earlier in relation to the Arltian technological dream, it is possible to
oppose to the simulations of the State a further set of simulations which confuse the codes which constitute what counts as real. The fake functions to reveal the fake, as explained by the museum guard, Fuyita, whose job it is to regulate the machine's output, but who has become seduced by this cyberpunk
version of Macedonio's *Museo de la novela de la Eterna*:

> —El poder político es siempre criminal —dijo Fuyita—. El Presidente es un loco,
> sus ministros son todos psicópatas. El Estado argentino es telépata, sus servicios de
> inteligencia captan la mente ajena. [... P]ero la [...] máquina ha logrado infiltrarse en
> sus redes, ya no distinguen la historia cierta de las versiones falsas. Existe una cierta
> relación entre la facultad telepática y la televisión [...] el ojo técnico-miope de la cámara

19   *Ibid.*, p. 60.

graba y transmite los pensamientos reprimidos y hostiles de las masas convertidas en imágenes. Ver televisión es leer el pensamiento de millones de personas.[21]

Specifically in relation to our theme of science and the creative imagination, Russo sums up Piglia's thinking — which takes us all the way back to Arlt — about the way science as discourse underpins the technocratic State and for that very reason becomes a key site of discursive resignification for the creative artist. Noting that it was physicists who gave the name 'quark' to their new elementary particle, in homage to Joyce's *Finnegan's Wake*, Russo comments:

Si los políticos les creen a los científicos y los científicos les creen a los novelistas, la conclusión era sencilla. Había que influir sobre la realidad y usar los métodos de la ciencia para inventar un mundo donde un soldado que se pasa treinta años metido en la selva obedeciendo órdenes sea imposible [...].[22]

Controlling the Code

I now turn to some of the attempted strategies of control over the proliferation of simulacra as represented in the texts. Both *La ciudad ausente* and *Hombre mirando al sudeste* suggest very strongly that idea, expressed forcefully also by Baudrillard, that what the State cannot ultimately tolerate is a simulation that reveals the 'democratic' State itself, in its very mode of operation, to be the prime generator of simulacra. As Doctor Denis puts it to Rantés in allegorical form, 'Si usted no es un chiflado, yo tendría que admitir que es realmente un extraterrestre. ¿Sabe lo que eso significaría? Que el chiflado soy yo.' This is the way in which Baudrillard puts the same idea, in relation to the supposed moral scandal and outrage at the revelations of Watergate:

… this is what must be said at all costs, for this is what everyone is concerned to conceal, this dissimulation masking a strengthening of morality, a moral panic as we approach the primal *(mise en) scène* of capital: its instantaneous cruelty, its incomprehensible ferocity, its fundamental immorality — this is what is scandalous.[23]

The primary images of attempted control in these texts are those of museums and clinics. A museum appears both in *La invención de Morel* and in

20   *Ibid.*, p. 144.
21   *Ibid.*, p. 63.
22   *Ibid.*, p. 142.
23   Baudrillard (1988), p. 173.

La ciudad ausente, while the image of the psychiatric clinic is mentioned in all three texts and is fundamental to Subiela and Piglia.

Museums

La ciudad ausente sums up the way in which the museum is used by the State as a way of controlling, of institutionally framing, the dangerous proliferation of stories which reveals the reality effect to be itself one more (powerful) simulation. Russo explains:

> *Hace quince años que cayó el Muro de Berlín y lo único que queda es la máquina y la memoria de la máquina y no hay otra cosa [...] nada, sólo el rastrojo, el campo seco, las marcas de la escarcha.*[24] *Por eso la quieren desactivar. Primero, cuando vieron que no la podían desconocer, cuando se supo que hasta los cuentos de Borges venían de la máquina de Macedonio, que incluso estaban circulando versiones nuevas sobre lo que había pasado en las Malvinas; entonces decidieron llevarla al Museo, inventarle un Museo [...] a ver si la podían anular, convertirla en lo que se llama una pieza de museo, un mundo muerto, pero las historias se reproducían por todos lados, no pudieron pararla, relatos y relatos y relatos.*[25]

The museum in *La invención de Morel* is somewhat different, but equally revealing in the fact that it occupies the same imaginative space as the virtual reality machine. Morel, when he first invented his recording and projecting machine, had intended to use it to build great living albums or museums, both public and private, with his images: 'La palabra *museo*, que uso para designar esta casa, es una sobrevivencia del tiempo en que trabajaba los proyectos de mi invento, sin conocimiento de su alcance'.[26] But this kind of framing of simulacra becomes somehow useless once the invention has so totally consumed the real that it is no longer possible to distinguish between the two, when, as we discover, even the sun and the moon have been recorded by the machine and are replicated in its projection. One is reminded of Baudrillard's 'outrageous' contention that even the 'sun' is no longer real.

Clinics

The same concept of the (attempted) institutional framing of forms of simulation applies to the clinic or asylum, which for Subiela and Piglia are

---

24   This reference, repeated throughout the novel, to the marks of frost revealing previously disturbed earth represents the almost invisible traces left by the anonymous graves of the disappeared.

25   Piglia (1993), pp. 144–5.

26   Bioy Casares ([1940] 1991), p. 93.

very clearly part of a mode of control and policing of the reality principle. Dr Denis in *Hombre mirando al sudeste* and Dr Aranda of the story of the 'Nudos blancos' in *La ciudad ausente* are both figured as complicit with the methods and the techniques of the torturer. Both Rantés and Elena, as inmates of their respective clinics, claim to be machines, and machines that cannot feel: '—¿Qué es ser una máquina? —preguntó el doctor Aranda. / —Nada —dijo ella—. Una máquina no es; una máquina funciona'.[27] Both are 'simuladores' and both claim that the best way to protect their mission would be to tell the truth: '¿Sabe cuál es la mejor manera de proteger mi misión? Decir la verdad,' claims Rantés, while Elena 'decidió que iba a decir la verdad. Era una loca que creía ser una mujer policía a la que obligaban a internarse en una clínica psiquiátrica y era una mujer policía entrenada para fingir que estaba en una máquina exhibida en la sala de un Museo.'[28] The point, as for Baudrillard, is that what counts as truth in this context is impossible to distinguish from the simulation: as would occur if you were to go into a bank and attempt to simulate a robbery.

A number of these ideas are brought together in one of the culminating sequences from *Hombre mirando al sudeste*, set in the pathology lab of the asylum where Rantés has been allowed to work as an assistant.[29] Rantés has gone one step too far in confronting the social contract, and the director of the asylum has intervened and ordered the doctor to sedate him. Rantés is standing alone at the pathology bench on which sits a human brain which he is weighing and analysing. In walks Doctor Denis, to tell Rantés that he will have to 'medicarlo' and that he will not be allowed to work as a lab assistant any more. Rantés, staring at the brain, comments:

> *Como archivan información lo entiendo. Pero ... ¿qué los hace funcionar, qué los mantiene en acción, qué los hace sentir? ¿Estará acá lo que ustedes llaman alma? [...] Hay torturadores que aman a Beethoven, quieren a sus hijos, van a misa ... El hombre se permite eso.*

As the doctor protests that it is all for Rantés' good (subsequently claiming, in an ironic echo of the Christian passion, that he will not abandon him), Rantés severs the brain with a large knife and, studying its internal contours, asks, '¿Dónde está aquella tarde en la que sintió por primera vez el amor de una mujer? ¿Qué marcas quedan … de los momentos de dolor o de goce que habrá sentido este hombre?' Taking the brain over to the

---

27   Piglia (1992), p. 68.
28   'Los nudos blancos', in Piglia (1992), p. 67.
29   The sequence can be viewed at http://www.latin-american.cam.ac.uk/simulacra/clips/.

sink, he tears off a piece and crumbles it under the running tap: 'Ahí va
Einstein, Bach, el señor Nadie, un loco, un asesino. ¿Usted qué cree, doctor, esta cloaca irá al cielo o al infierno?' The imagery and dialogue here
present us with an intensely moving allegory of the dissolution of the organic within a ferocious regime of technocratic simulation of the real. Not
coincidentally, this powerful allegory is coupled with the waning of those
traditional scientific discourses which produced the organic body as an object of knowledge in the first place.

La invención de Morel, for its part, makes, I believe, an indirect reference
to the lunatic asylum of The Cabinet of Dr Calligari:

Anoche soñé esto:

Yo estaba en un manicomio. Después de una larga consulta (¿el proceso?) con un
médico, mi familia me había llevado ahí. Morel era el director. Por momentos, yo
sabía que estaba en la isla; por momentos, creía estar en el manicomio; por momentos, era el director del manicomio.[30]

It is no accident that the ending of Dr Calligari, in which the hierarchy of
reason and madness is firmly put back into place, was added on by the production studio in order to neutralise the unsettling political message of the
film, with its suggestion of the madness that is constitutive of authority.

Conclusion: Return of the Foundational

At the end of La invención de Morel, as the narrator is detachedly observing
his own body decay before him — the price for the eternal life he believes
he will gain in the machine's virtual reality — the political re-emerges, and
specifically the politics of the Nation. In an apostrophe to his Patria, to
Venezuela, to the dream of Bolívar, the protagonist cries 'tú eres, Patria,
los señores del gobierno, las milicias con uniforme [...]; sin embargo te
quiero, y desde mi disolución muchas veces te saludo; [...] la emoción pa-
triótica, la emoción que ahora no reprimo'F.[31] How can we account for this
return of the foundational at the very moment of its dissolution into simulacrum, 'desde mi disolución,' as he says?

I believe this to be a very powerful metaphor for the way in which the
order of simulation announced in all of these texts resurrects myths of
origin, authenticity, 'lived' experience. As Baudrillard puts it, simulation is
a panic-stricken production of the real. To put it in more recent terms,
fundamentalism is the inevitable corollary of globalisation: residual lo-

30   Bioy Casares ([1940] 1991), p. 65.
31   Ibid., p. 126.

calisms — nationalism, regionalism, specificity, traditionalism, or some other foundational identity politics — are in some sense constructed by the very dissolutions which these texts chart.

As to the effectiveness of the strategies of resistance, of the shuffling of codes, evoked by Subiela and Piglia, in particular the latter, one would have to ask, in the vein of Terry Eagleton's critique of Lyotard, whether they merely amount in effect to 'an anarchist version of that very same epistemology, namely the guerrilla skirmishes of a 'paralogism' which might from time to time induce ruptures, instabilities, paradoxes and micro-catastrophic discontinuities into this terroristic techno-scientific system',[32] but which ultimately leave it intact.

---

32   Terry Eagleton (1986), 'Capitalism, Modernism and Postmodernism', in *Against the Grain: Selected Essays* (London: Verso), 131–47, quote from p. 134.

11

Desert Poetics of Mario Montalbetti: Writing, Knowledge, Topologies

William Rowe

> 'Art does not illustrate philosophy,
> *it comes before it.*' (Jeff Nuttall)

The desert is not a place of linear orientation. Borges's brief parable of 'Los dos reyes y los dos laberintos' makes this clear. When the King of Arabia wishes to take revenge on the King of Babilonia he has no need of architecture: after three days' journey bound to the back of a swift camel, he releases him into the desert. The desert makes an end of form: 'la escala termina con la forma', as Mario Montalbetti writes in *Fin desierto*.[1] Montalbetti's book explores what happens when lines become subject to the desert:

> ... *cada línea contiene su propia ausencia*
> *porque cada línea no importa*
> *la escala termina con la forma*
> *los ritmos y las texturas se desbandan sobre las dunas*[2]

That difference of scale is taken up in the typography of the first, screen-fold, edition of the book, which consists of a single sheet of yellowish paper several metres long and uses letters of different sizes, intensities and colours surrounded by a great deal of blank space. Mallarmé, in his preface to *Un coup de des*, points out that he has merely redistributed the white space at the margins of the poem. The effect is to bring what's outside the poem inside, not as thematics but as the borders of language, and to project the inside outside.

---

1 Mario Montalbetti, *Fin desierto*, p. 10. The first, screenfold, edition was published in 1995 by Studio A Editores, Lima. All quotations in this essay are from the second edition (Lima: Mosca Azul, 1997), which has the form of a conventional book.
2 *Ibid.*

The writing is not concerned with an *experience* of the desert: 'Más que experiencias, son ciertas propiedades (casi arquitectónicas) del desierto las que trato que se cuelen en el libro. Por ejemplo el minimalismo, no sólo en el sentido físico de que en un desierto hay pocas cosas, sino también en el sentido conceptual de que hay pocos tipos de cosas.'[3] The minimalism has a Wenders touch, as in Ry Cooder's sound track of blues chords in minimalist mode for the film 'Paris Texas'. There's a dispersed and minimal narrative of abandonment by someone, and a subsequent search for them:

> *este es el verso en que entro al pueblo*

> *y pregunto por ella y por un bar llamado el patio*
> *todos volteamos hacia el mismo lugar todos*
> *[cometimos el mismo error*
> *caminé*

> *por estos versos para olvidar tormentos y sentí un*
> *[alivio pasajero al ver*
> *jacarandás en flor*

> *pero luego todo volvió de golpe y no pude sino escupir*
> *[sobre estas calles*[4]

Crossing the desert and writing become the same time-place, not in the sense that the one is represented by the other, but because the conditions of cognition are prolonged from the one to the other. Without ceasing to resonate as a metaphor, crossing the desert becomes a condition of writing: to traverse the yellowish and largely empty space of the screenfold edition of the book is to cross an epistemological desert.

The relationship between space and cognition recurs as tone in the enunciation of concepts, such as *reason* and *telos*: 'No tengo razón ni fin', isolated from speech contexts, could be heard in a variety of tones. But emplaced in the book/desert, the terms become charged with loss and damage:

> *he dormido mal sobre almohadas envenenadas*
> *he dejado el sitio para siempre*

---

3   Daniel Salas Díaz (1996), 'Mario Montalbetti del desierto', *Somos* (Lima, April), p. 47.
4   Montalbetti (1997), p. 16.

no tengo razón ni fin soy paisaje mínimo llevo los dientes
[rotos[5]

The tone of enunciation is inseparable from the concepts themselves. This is poetry, not philosophy or science. Does that then mean that the concepts are a bit less serious, and/or the capability for exactness of language diminished? What does the answer depend upon? Partly, for sure, on the current relationship between science and the humanities, i.e. as conditioning the way in which one reads.

It also depends on language, i.e. on the scene of language which the book constructs, but also too on what is available as evidence. Different disciplines work with different definitions of evidence: the 'facts' in literary studies are not the same as the 'facts' in science. This is one of the main reasons why it is difficult to study literature and science as parts of a single field. It is not sufficient though to stop at the differences between what is taken to constitute evidence. One has to investigate how a particular delimitation is itself constituted. The Peruvian philosopher, José Carlos Ballón, in his book *Un cambio en nuestro paradigma de ciencia*, proposes the term 'compromiso ontológico' as a way of gathering together the various dimensions of the problem:

> *Entiendo por compromiso ontológico aquellos 'compromisos básicos' o 'profesionales'*
> *— para usar inicialmente la expresión de Thomas Kuhn — mediante los cuales*
> *todas las disciplinas científicas 'especifican qué tipo de entidades contiene el universo,*
> *y por implicación, aquellas que no contiene'. Es decir, aquellos que nos permiten de-*
> *cidir por razones que no son de origen experimental, qué es lo que existe y qué es lo*
> *que no existe.*[6]

In this chapter I will suggest some of the ways in which Montalbetti's work displays an 'ontological commitment' which intersects with Ballón's epistemological concern with contemporary science. I will also speculate about how their work can be read as a regional, i.e. Peruvian, intervention in the field.

The scene of enunciation is the place which makes evident what is taken to be evidence. The Latin root *videre* in the word evidence should not lead us to assume that what is at stake is merely visual perception. The

---

5   *Ibid.*, p.32.
6   José Carlos Ballón (1999) *Un cambio en nuestro paradigma de ciencia* (Lima: Universidad Nacional Mayor de San Marcos), p. 237.

place which makes evident is actually an assemblage of enunciation,[7] as well as a gathering or assemblage of things. It is a place of inscription as well as being the result of a practice of knowledge. Consider Montalbetti's use of the shaman's *mesa* (= table and mass/*misa*):

> *sobre la mesa hay animales vivos y flores amarillas de*
> *[montaña*
>
> *muertes simples que se clavan en la tierra como estacas de*
> *[plata*
> *estampas de los santos gregorio santiago y benedicto*
>
> *la luna vacía y el sol de invierno*
>
> *[…]*
> *todo está sobre la mesa*
> *sobre la mesa las hojas de coca y los nevados*
> *y los ríos de obsidiana*[8]

The *mesa/misa* (the vowels are indistinguishable in Quechua) is not universalised as folklore or myth. It is taken to be a practice of knowledge, a place of inscription and a definition of evidence: therefore a place for place. The coca leaves, the snow peaks and the subsequent mention of a condor are then permitted to inscribe this as a Peruvian place, these being not just regional things but sacred objects in Andean religion, just as the Catholic saints are indices of its synergies.

On the other hand, there is no question of the regional as value, only of exposing how the relation between place and language has become residual in an epoch of globalised languages. Montalbetti in addition takes the shaman's table as a book before writing,[9] an action which not only fol-

---

7 See Gilles Deleuze and Felix Guattari, *A Thousand Plateaus*: an assemblage of enunciation is where 'the non-discursive' presuppositions of a language action are to be found; there is no individual enunciation' (pp. 78–9). Note that this overlaps with Wittgenstein's idea of language-games, such that in both the production of evidence is a social action.

8 Montalbetti (1997), p. 11.

9 See Jerome Rothenberg (2000) 'The Poetics and Ethnopoetics of the Book and Writing,' in Jerome Rothenberg and Stephen Clay (eds), *A Book of the Book: Some Works and Projections about the Book and Writing* (New York: Granary Books), pp. 7–16. See also the work of Joseph Beuys, where the shaman's *mesa* or altar is used as an interface between sculpture, book and performance.

lows through from the European avant-gardes' uses of the primitive but also removes the letter in Peru from its history of colonial imposition.[10] Thus Montalbetti's book displays the genealogy of its own production as writing, placing the latter in relation to the production of knowledge in Peru. This in turn includes a series of modes of the gathering of data, such as a museum, the book as a museum:

> *hay un museo que contiene réplicas*
> *de todo lo que has oído*
> *hay un libro que repite todo lo que escribes*
> *y otro que escribe todo lo que repites*[11]

but the museum is relocated, against the grain of the Republican episteme, alongside a different logic of space:

> *hay un sol partido en dos*
> *y una sombra espesa en la escisión*

which places scision before us as beginning of knowledge, in the manner of Hesiod.[12] The divided sun as cosmogonic and logogonic gap is antecedent to the lines which disappear in the desert.

Anthropology is the discipline which turns shamanism into evidence. Its debt to the Enlightenment (and Republican) episteme is indicated in the commitment to explanation. It is an intellectual practice whose foundational claims to being a discipline rest upon the capacity to explain phenomena like shamanism. That depends upon gathering information from informants via the typical question '¿por qué lo hacen de esa manera?'[13] However, the repeated answer elicited ('así lo hacen así lo hacemos') stymies any explanation in other terms than those of the practitioners in their practice. The difficulty is not just one of terms, but of language and poetics.[14]

---

10 To do that is to go against the *Indigenista* narrative of Peruvian literary history, which takes orality to be the site of resistance to the colonial inheritance. For a recent example of that narrative see Antonio Cornejo Polar's (1994) *Escribir en el aire* (Lima: Editorial Horizonte).

11 Montalbetti (1997), p. 10.

12 Edward S. Casey (1998) *The Fate of Place* (Berkely: University of California Press), p. 16.

13 Montalbetti (1997), p. 12.

14 See James Clifford (1986) *The Poetics and Politics of Ethnography* (Berkely: University of California Press).

Yet Montalbetti does not stop at that point, which corresponds, roughly, to a radical empiricism. He moves into the desert, which is a different
place of writing from the Shaman's *mesa* and which alters the relationship
between words and things: the fact that there are not just few things but
few types of thing changes the epistemological environment:

> *todos lo saben todos lo han visto*
> *y están todos ciegos de ver tanta ausencia*

The recurrent statements of absence or emptiness have to do not with
Buddhism but with the limitations of any merely phenomenological view
of the environment. Montalbetti's book, in one of its key actions, uses the
desert as a counter-environment, in order, by contrast, to render visible the
current globalised world where there are more and more images and less
and less to perceive. To immerse oneself in the desert is to engage in the
counterinductive strategy recommended by Paul Feyerabend:

> how can we possibly examine something we are using all the time?
> How can we analyse the terms in which we habitually express our most
> simple and straightforward observations [...] How can we discover the
> kind of world we presuppose when proceeding as we do?

> [...] we cannot discover it from the *inside*. We need an *external* standard
> of criticism [...] an entire alternative world. [...] We must invent a
> new conceptual system that suspends, or clashes with the most care-
> fully established observational results, confounds the most plausible
> theoretical principles, and introduces perceptions that cannot form
> part of the existing perceptual world.[15]

To enter the desert is therefore a necessity — or counter-necessity, against
the necessities imposed by globalisation.

Yet any approximation to Montalbetti's book which tries to speak it
from an inside, i.e. to make it into a habitable inside, runs the risk of di-
minishing its engagement with a multiplicity of outsides, which are no less
'present'. To say that something is simply 'in' a poem or 'in' a text, a habit
which the procedure of close reading has tended to encourage, is a case of
what Whitehead calls 'simple location'.

---

15   Paul Feyerabend (1988) *Against Method* (London: Verso), pp. 22–3.

As soon as you have settled, however you do settle, what you mean by *a definite place* in space-time, you can adequately state the relation of a particular material body to space-time by saying that it is just there, *in that place*; and, so far as simple location is concerned, there is nothing more to be said on the subject.[16]

The various events that occur as the poem-reading, that make up what people call the text, are complex in that they gather what's before, after, and elsewhere, in what A.N. Whitehead calls 'the prehensive unification of surrounding spaces'[17] that is effectuated by the body. One of the many outsides which the poem involves is the location of the dead and their traces, a region visited by epic poems such as Juan L. Ortiz's *El Gualeguay*, or Raúl Zurita's *Anteparaíso*, or — why not? — *Pedro Páramo*.

It's the tone of *Fin desierto* that has an identificatory effect: by generating a sense of voice, and of a person speaking, it invites the feeling of an inhabitable inside. That sense of an 'inside the book' has affinities with the image of reading delineated by early modernists such as Rilke and Proust: a room in a house or hotel, i.e. an interior architectonic space which one exits from when one closes the book and walks out of the room into the wider space of the street But the shaman's *mesa*, as model of reading, proposes a considerably more extensive process of location:

> sobre la mesa las tormentas y los vientos y los lagos
> de altura
> la sed continua de las gargantas en las islas[18]

And, more pervasively, the desert as location infiltrates the writing of the book.

Yet the desert does not offer a place for place. If the tone gives an inside, it's immediately confronted by a conceptual homelessness, which is not the same as a concept of homelessness. Theodor Adorno's *Negative Dialectics* offers an example of tone in relation to concepts, not just outside them but inside them, or moving from one to the other. When tone moves into the concept it becomes something else, no longer a voice but the enunciation of what concepts have not grasped, the basic tenor of *Negative Dialectics* being the need for philosophy to confront all that which has been

16   Quoted by Casey (1998), p. 211.
17   *Ibid.*, p. 214. In *Process and Reality* (New York, Free Press, 1979), Whitehead writes: 'actual entities involve each other by reason of their prehensions of each other' (p. 20).
18   Montalbetti (1997), p. 12.

excluded from its concepts. Adorno writes for example, 'objects do not go into their concepts without leaving a remainder [...] the concept does not exhaust the thing conceived', one of the book's motivations being 'to transcend the official separation' of philosophy and science. And of philosophy and art: 'no object is wholly known; knowledge is not supposed to prepare the phantasm of a whole. Thus the goal of a philosophical interpretation of works [of art] cannot be their identification with the concept.' And with all this, 'no reforms within the world sufficed to do justice to the dead'.[19] The last statement is a paraphrase of Kant, but it also serves to express Adorno's concern with Auschwitz.

Montalbetti's work not only brings into awareness what concepts leave out, but also uses the desert scenario to deprive concepts of the territory which justifies them by making them operative. The lines which disappear into the desert bring these themes together:

> *Cada línea es un río una calle un color imaginario*
> *un número irracional en medio de una suma infrecuente*
> *el rostro cambiante de una ventana un amanecer en tu boca*
> *una lápida una lápida que no coagula...*

> *porque cada línea contiene su propia ausencia*
> *porque cada línea no importa*

> *la escala termina con la forma*[20]

With Adorno the movement is from concepts to what they can't contain back to the necessity for concepts (i.e. for philosophy). With Montalbetti we go from the arising of concepts to an absence that haunts all of them. For Adorno, 'the course of history forces materialism upon metaphysics, traditionally the direct antithesis of materialism' — materialism as in 'the somatic, unmeaningful stratum of life [which] is the stage of suffering', of the suffering of the camps for example. 'Children sense some of this [writes Adorno] in the fascination that issues from the flayer's zone, from carcasses, from the repulsively sweet odor of putrefaction.'[21] And for Montalbetti,

19   Theodor Adorno (1990), *Negative Dialectics* (London: Routledge), pp. 5, xx, 14, 385
20   Montalbetti (1997), p. 10.
21   Adorno (1990), pp. 365–6.

he aprendido en todo esto a no mirar
con desprecio al virus o al verano
porque también ellos de incomprensible manera

armonizan con todo lo que calla y así se expresa

el olor de los cadáveres
que es perfume de ángeles
ha obligado a cerrar el aeropuerto
ya no viene el que viene y no es el que es[22]

The memory elicited, for a Peruvian perspective, is Ayacucho airport,[23] but it can include the victims of other post-Auschwitz genocidal actions, as well as after-images of the holocaust:

caen en sucesión uno tras otra
embellecidos por los tatuajes de kaposi[24]
tras otra tras uno

raspando del aire oxígenos letales y derramando

de sus labios una emulsión de plata
que revela sus cuerpos contra oscuras cámaras
que los devuelven pálidos[25]

The photograph as death-trace, the last breath as graphism, traces as sarkographic.[26]

The sight of human remains in the desert is a recurrent scene in *Fin desierto*. In the work of Antonio Cisneros, the remains of burials in the Peruvian desert disturb the grand narratives of hispanism (glorious conquest) and of indigenism (native identity); their image gives the lie to the former and is insufficient for the latter:

---

22 Montalbetti (1997), p. 15.
23 More people were disappeared in the Department of Ayacucho in the decade of 1980 than in Chile under Pinochet.
24 Kaposi's sarcoma: 'multiple haemorrhagic sarcomata, especially affecting the skin of the extremities' (*Butterworth's Medical Dictionary*, London: Butterworths, 1978, p. 930).
25 Montalbetti (1997), p. 15
26 'Cámaras' can also be read as gas chambers.

Sólo trapos
y cráneos de los muertos, nos anuncian

que bajo estas arenas
sembraron en manada a nuestros padres.

For Rodrigo Quijano, writing in the 1990s, the southern desert, where the best beaches are, is also a key scenario:

puras formas de avanzar enmascarado entre el cabello más
[oscuro de la madrugada
donde los huesos eran marcas de una historia interesante
pero de una geografía estéril, que advertía a la conciencia,
mientras yo avanzaba hacia las playas listo para
[zambullirme
en arqueológica bondad
pues eran los reinos del topo, invisibles para el reino de los
[cielos
pues eran las tumbas de pasto, donde reposaban mis
[antepasados[27]

History has lost its resonance, but the question of identity still presses, though 'archaeological happiness' mitigates the story of identity with critical irony. But Montalbetti's book does not entertain any discourse of identity, of 'padres' and 'antepasados'. The desert is Peruvian, but it is also elsewhere, anywhere where the earth yields scraps of human beings, traces of a condition which includes both pain and genocide ('se está mucha gente por lo visto / enterrada de noche en la arena').[28] The recurrent scene of people, bones, craniums buried in the sand is not located in an already existing geography, it is itself locatory. It locates phenomena that are global, against globalisation. Archaeology not only no longer feeds indigenism and nationalism, it disinters the crimes of the victors in the age of globalisation.[29]

---

27   Rodrigo Quijano (1998) *Una procesión entera va por dentro* (Lima: Ritual de lo Habitual Ediciones), p. 19.

28   Montalbetti (1997), p. 30.

29   For a discussion of how difficult that might be after the Gulf War, and subsequent wars carried out by the so-called 'international community', see William Rowe (2001) *Memoria, continuidad, multitemporalidad*, Cuadernos de Literatura, 37 (La Paz: Universidad Mayor de San Andrés).

What archaeology uncovers intersects with the other which is inside love:

> *viven en la oscuridad durante el amor*
> *y están y no están en sucesión*
> *[...]*
>
> *nada es el amor durante el amor*
>
> *excavaron toda la mañana*
> *hoyos circulares junto a osamentas de buitres incompletos*
> *y también ahí encontraron los jóvenes arqueólogos*
>
> *el íntimo intercambio de heridas y de besos*
> *lavados por la arena devastada por la arena*
>
> *nada es el amor durante el amor*[30]

The archaeological event is placed beside the other in time, in matter, in pain, in love.

Another form of disinterment, subsequent to the burial in the 1970s of Yungay, a town in the Peruvian central highlands under several metres of mud, as a result of an earthquake, is placed beside the shattering of the Word:

> *la palabra ha sido quebrantada*
> *y la suma de todos sus fragmentos*
>
> *es ahora destrucción*
>
> *hay flores hay cavidades craneanas*
> *mástiles parecidos a los camiones*
> *exhumados en yungay palmeras de barro*[31]

The broken word, disinterred suffering and that which is not in concepts intersect. But which inheritance of concepts? For José Carlos Ballón, the difference between Newtonian physics and the twentieth-century scientific revolutions is not so much to do with the results of experiments, ob-

---

30   Montalbetti (1997), p. 36.
31   *Ibid.*, p. 18.

servation and new theories invented to explain them, which is the usual narrative, as with a different 'compromiso ontológico' which involves questions of representation as much as it does what goes on in the laboratory. To sketch out his argument, the Newtonian belief in direct and transparent representations of nature depends upon an 'atomist-reductionist' theory of meaning, which reaches its fullest expression in the early Wittgenstein or in Russell,[32] which has the effect of turning theories into propositions that can be tested statement by statement against the 'facts'. There are similarities here with Feyerabend's arguments in *Against Method*, but what is particular to Ballón is to join the critique of representation in science with a criticism of language, and to place both at the disposal of a revision of modernity from a Peruvian perspective. He finds the clearest formulation of the 'tesis gnoseológica fundante del discurso filosófico de la modernidad' in Descartes' method. The Cartesian principle of the clarity and distinctness of an idea 'consiste en la *completa determinación de su significado* (lo que necesariamente es), despojada de sus relaciones con todo lo demás (lo que no es [...]). En ello consiste llegar a su naturaleza simple, y tal es la base de su certeza puntual.'[33] In Whitehead's terms, that would be a case of simple location. Among the Cartesian principles which underwrite the correspondence of concepts and objects, there are, as Ballón points out, two key ideas: 1) that we cannot doubt the continuity of a physical body, in other words that there can be no gaps in its extension, otherwise one would be involved in believing in 'la existencia de la nada'. And 2) that we cannot doubt the existence of thought. Putting the two together corresponds to the soceity of 'selfish individualism', to use MacPherson's term, or what Ballón calls the fetishism of 'el individuo aislado'.

Ballón gives great importance to the distinction between modern, i.e. classical, science and 'the contemporary episteme', which he defines, for example, through the notion that 'objectivity is understood as social intersubjectivity'. If Newtonian science prospered because of historical and cultural conditions, they are conditions which do not pertain in Peru, where

> ...*inmensas 'dificultades individuales' y 'obstáculos institucionales' [...] tienen que salvarse [...] para 'producir un régimen estable de acumulación científica. [...] Para que la ciencia arraigue en la sociedad, es preciso que los intereses de sus agentes (tanto intelectuales como económicos y políticos) coincidan en el apoyo al estudio racional de la naturaleza.'*

32   Ballón (1999, pp. 264–5.
33   *Ibid.*, p. 266.

But the argument can be turned round: if the scientific revolutions of the twentieth century are unfinished revolutions, as Ballón argues,[34] then in Peru, which never fitted the 'compromiso ontológico' of modernity, might they not have more chance of prospering? The question is mine, not Ballón's, but I think it is implicit in his book.

The distance between Montalbetti's poetics and those of the founders of Modernism is shown among other things by his definition of beauty:

> *los cantos*
> *que canté para nadie ahora son tuyos*
>
> *porque no puedes ser lo que eres / amor*
> *porque no puedes sentir lo que sientes*
>
> *sin la incomparable belleza de lo que no eres*
> *de lo que no sientes*
>
> *asómate al borde de tu corazón y observa*
> *la inmunda danza de las neoplasias*
> *festejando la debacle de las oraciones en ese lugar*[35]

In the first edition 'los cantos' is printed in bold and in larger type, in its own space to the right of the rest of the text. The figuration of beauty in Pound's *Cantos* is without residue, or disease, or rubbish.[36] The heart, before it became an everyday metaphor, was a spatialisation of emotion. Montalbetti's concern is with what the word leaves out: 'toda esta basura que viene con ser humano'.[37] *Corazón*, as representation of affect, makes affect into sentiment, by a process of exclusion (sentiment is always a response to less than the full picture), just as the image of the beloved excludes cancer. Montalbetti turns the relationship round: to criticise the image by what it excludes — and that includes the smell of corpses. What is not in representations of affect, and what is not in concepts, overlap in what is not in the word.

---

34   Ballón (1999), pp. 415–6, 284, 262–3, 262 notes 356, 136.
35   Montalbetti (1997), p. 14.
36   See Out To Lunch, 'Garbage — a Discussion of Value,' *PORES*, vol. 1, October 2001 (www.pores.bbk.ac.uk).
37   Montalbetti (1997), p. 29.

Certain of Michel Serres' statements, in his conversations with Bruno Latour, can serve to illuminate the relationship between Montalbetti's work and Ballón's: *'knowledge* and *misfortune* cannot be separated; each is as objective and no doubt universal as the other. By only knowing or living one of them, we are unaware of what we think, what we do and what we are.'[38] This insistence on the relationship of knowledge and suffering is a challenge to rethink the role of the Humanities. What is denied by the common idea of science — the fact that 'our capacities come from our weaknesses' — has produced 'the forgetting of the humanities — that continuous cry of suffering, that multiple and universal expression, in every language, of human misfortune'.[39]

Montalbetti works with classical space, not quantum, or, more accurately, takes reading to an edge where classical space starts to break down. Ballón, on the other hand, elaborates out of quantum physics the possibilities for a different notion of knowledge in the future. Montalbetti explores the consequences for language, and Ballón those that affect the epistemological culture. Both show the representational contract associated with modern, i.e. classical science, becoming inoperative — which doesn't prevent it still being used. Montalbetti gets there via pain and absence, and Ballón via a change of 'ontological commitment' with its social, cultural and not just scientific consequences. Both reunite knowledge with the whole human context.

The densest and perhaps most far-reaching of Montalbetti's reflections have to do with language. Auschwitz infiltrates the critique of language, and both of them the theory of writing:

atravesamos el tiempo

mi lenguaje no es de este mundo
pero mis palabras sí lo son

aquí están las llagas de mis encías[40]

The criticism of language borders on a wholly Other, a territory similar to that which Paul Celan explored in his later poetry. The sentence 'aquí hubo

---

38 Michel Serres (1995) *Conversations on Science, Culture, and Time* (Ann Arbor: University of Michigan Press), p. 181.
39 *Ibid.*, pp. 182, 180.
40 Montalbetti (1997), p. 28.

alguien que rompió su palabra' takes on a different accent, other than its historical ambience:

> *aquí hubo alguien que rompió su palabra*
> *[...]*
>
> *lo que dura cruzar el desierto*
> *lo que dura cruzar la palabra*
>
> *torah tora torah*

Tora is the Word of God, revealed to Moses when crossing the desert, and so it is the word for the word.

Montalbetti's book sets out the elements of a theory of writing which extends the critique of language carried out by the early twentieth-century avant gardes. This theory places actual — rather than transcendental — time and space inside the word, but it also asks in what space are words themselves located:

> *las palabras que son como pozos que contienen su propia*
> *[ausencia*
> *¿dónde están?*
>
> *entre las letras en los espacios ciegos en la fruta*
> *[picada*
> *pero también*
> *en el ojo de la orca en la boca de la hostia en la carne*
> *[acecinada*[41]

The capability of words to contain space is here inseparable from their absence, an absence which they themselves make. Therefore words cannot be assimilated to substance or matter or any other concept of material extension. They are always unaccommodatable, except to a desert condition. But words also exist as phonological mass, whose movements carry history ('hay un desierto a la deriva' is the first line of the book), a drift inside which genocide can be heard — *carne acecinada*, in Latin American pronunciation, is both bacon and murdered flesh.

---

41   *Ibid.*, p. 58.

Absence is both inside the word and outside it. Topologically, a hole that contains its own absence is a hole characterised both by having emptiness inside it and by enclosing its own making of emptiness — 'the word for word is word', as William Burroughs wrote. The topogenetic effect gets called 'magnetic suction', in lines which are also a homage to Borges:

> *aquí hay alguien*
> *que se ha ido y que ha dejado*
> *esta succión imantada*
> *y que piensa por nosotros*
> *desde el fondo de un espejo*[42]

Another passage enacts a similar movement but without naming that 'someone' who haunts the word:

> *[...] el lugar es la oscuridad*
> *entre los destellos*
> *paraje de sauces y ríos*
>
> *un manto sobre cada cosa*[43]

Manto is defined by the dictionary as 'velo; capa que se pone sobre las imágenes de la virgen; capa grasienta en que nacen envueltas las criaturas'. Place: a darkness between light sources, a quasi-skin that envelopes things, and not a site in an abstract, geometrical and global grid. Does the desert both allow no place for place and, by a folding of time, bring us to a past and future conjuncture where topogenesis becomes possible again?

Walter Benjamin, in an early essay, rejects the referential theory of language (it's an expression of bourgeois accumulation, he says) and asserts a view which is similar to Montalbetti's:

Language communicates the linguistic being of things. The clearest manifestation of this being, however, is language itself. The answer to the question 'What does language communicate?' is therefore 'All language communicates itself. [...] The language of this lamp, for example, does not communicate the lamp [...] but: the language-lamp, the lamp in communication, the lamp in expression.'[44]

---

42   *Ibid.*, p. 13.
43   *Ibid.*, p. 52.
44   Walter Benjamin (1984) *One Way Street, and Other Wrtitings* (London: Verso), p. 109.

If for Montalbetti the word contains its own making of absence, in Benjamin's essay language contains language, the word contains the name. The similarity is not only topological: Benjamin is also grappling with what is 'mystical' or 'inexplicable' in language: 'all language communicates itself *in* itself [...] if one chooses to call this immediacy magic, then the primary problem of language is its magic.' He uses book of Genesis as an initial setting out of the problem:

> In this 'Let there be' and in the words 'He named' at the beginning and end of the act, the deep and clear relation of   the creative act to language appears each time. [...] Language is therefore both creative and the finished creation, it is word and name. In God name is creative because it is word, and God's word is cognizant because it is name.[45]

Benjamin finds in the Judaic notion of the language of God an adequate mise en scène of the theory of language. Montalbetti's mis en scène is the desert phase of space-time. He adds something which is not in Benjamin and which belongs to the later twentieth century: a sense of the destruction of language ('todos los nombres han muerto', he writes,[46] echoing Huidobro's *Altazor*), and a concomitant sense of coming to a threshold which is not simply the silence between acts of language but which is the thought of abandoning language altogether as in the work of William Burroughs. What Wittgenstein says in Part 1 of the *Philosophical Investigations* — 'I must not saw off the branch on which I am sitting'[47] — can also be read as a recognition of the desire to do just that.

'Aquí no hay nada para la puesta en escena,' Montalbetti writes, 'porque todo fue puesto en abismo',[48] in a movement from space as container to what contains the container: 'todo este espacio y ningún lugar para ponerlo'.[49] The question of this other, this 'ningún lugar', is also figured in a criticism of the phenomenological account of seeing:

> *[...] no hay ojo que vea*
> *propiamente*

---

45   *Ibid.*, p. 114.
46   Montalbetti (1997), p. 34.
47   Ludwig Wittgenstein, (1967) *Philosophical Investigations* (Oxford: Blackwell), p. 27.
48   Montalbetti (1997), p. 32.
49   *Ibid.*, p. 20.

porque es aguja y agujero al mismo tiempo
el mismo nervio
óptico[50]

For Merleau-Ponty, place is produced by the body: 'the perceptual synthesis must be accomplished by the subject [...] This subject, which takes a point of view, is my body as the field of perception and action.'[51] But if the optic nerve is aprobe and an emptiness, an *agujero*, at the same time, then the perceiving subject disappears. Because we have eyes we say we see, but that does not mean there is something there to see:

ya que tenemos ojos
suponemos que hay algo que ver

pero no hay nada que ver

o lo que tenemos que ver
no se ve con los ojos[52]

Heaven is configured in the relationship of nail to hole. The nail is to the hole as the hole is to heaven, but in a hierarchy of nothing:

si quieres ganar el cielo primero debes saber perderlo

recoge por ejemplo un clavo
e imagina el agujero del que provino
[...]

arroja el clavo
guarda el agujero

arroja el agujero al suelo[53]

---

50   *Ibid.*, p. 19.
51   Merleau-Ponty (1964) *The Primacy of Perception* (Chicago: Northwestern University Press), p. 16.
52   Montalbetti (1997), p. 57.
53   *Ibid.*, p. 56.

Absence is also a multiplicity that becomes singular, as in a singular encounter, in Paul Celan's sense of an encounter with a 'wholly Other' as
something a poem might bring about:[54]

> *todos los nombres han muerto*
> *menos aquél que se ha ido dice*
> *[...]*
>
> *miente ahora*
> *di que has visto su rostro*
> *y que sigues vivo*
>
> *o que vives solo*[55]

To say that a meeting with an angel haunts Montalbetti's writing would
not be entirely accurate unless one added that that meeting is the other
side of the destruction of language. The desert in and of his writing has
similar spatial characteristics to the realm of active intelligence explored
by the Sufi visionary recitals called *ta'wil*. Both are a place of the arising
and loss of form (lo que adquiere forma / está condenado / a perderla),[56]
located between the sensible and the intelligible. The orientation of the
visionary recitals is towards the encounter with celestial beings, of whom
Avicenna — scientist, philosopher and theologian — writes, in one of his
recitals, 'they have been endowed with a shining aspect, a beauty that sets
the beholder trembling with admiration [...] they all live in the desert;
they have no need of dwelling places or shelter. To reach that place one
needs first to drink of the 'Spring of Life', which enables one 'to cross
vast deserts'. The place over which the water flows, and the world which
drinking it brings about, is an 'interval extending between the intelligible
and the sensible.'[57] Henri Corbin, from whom I have been quoting, trans-
lates this zone as the region of 'active imagination', which can be taken as
a synonym of the creative imagination. But the encounter in Montalbetti
is also with emptiness:

---

54 John Felstiner (2001), *Paul Celan: Poet, Survivor, Jew* (New Haven: Yale University
    Press), p. 165.
55 Montalbetti (1997), p. 34.
56 *Ibid.*, p. 23.
57 Henri Corbin (1990) *Avicenna and the Visionary Recital* (New Jersey: Princeton
    University Press), pp. 149, 141, 161.

siente el abismo entre
tus brazos circula libre
entre turbulencias aprecia
las gravedades y medita
en aquello que te idea [58]

This encounter with nothing is an occasion of astonishing beauty:

nunca ve nada el ciego
ni nada escucha siempre el sordo
no hay desierto sin nada
ni aroma absolutamente perdido

cuando habla la rosa antigua
y ora y muerde y extraña
y predice el parco pantano
en el que el eco se esconde

y desnuda se enreda la noche encima
algo vigila el oso cuando vigila
las constelaciones ausentes

y sólo hay totalmente nada
en la bulla de las lenguas [59]

---

58 Montalbetti (1997), p. 13. Turbulence is a term used by Michel Serres in order to gain access to a way of thinking which does not organise 'the totality of everything' with concepts like "'ideas', or 'categories', with references to the 'knowing subject', 'the analysis of language', and so on" (1995, p. 112).

59 Montalbetti (1997), p. 51.

Printed in the United Kingdom
by Lightning Source UK Ltd.
122418UK00002B/271/A